SLIME

Slime

A Natural History

Susanne Wedlich

Translated from German by Ayça Türkoğlu

GRANTA

Granta Publications, 12 Addison Avenue, London W11 4QR
First published in Great Britain by Granta Books, 2021

First published in German as *Das Buch vom Schleim* in 2019, the Naturkunden series, ed. by Judith Schalansky at Matthes & Seitz Berlin

The translation of this work was supported by a grant from the Goethe-Institut.

A CIP catalogue record for this book is available from the British Library.

1 3 5 7 9 10 8 6 4 2

ISBN 978 1 78378 670 1
eISBN 978 1 78378 684 8

Typeset in BemboStd by Avon DataSet Ltd, Arden Court, Alcester, Warwickshire B49 6HN
Printed and bound by CPI Group (UK) Ltd, Croydon, CR0 4YY
www.granta.com

For Noah, Elias and Roland

Contents

Introduction

Run – Don't walk from the Blob!
It crawls! It creeps! It eats you alive!
Trailer for *The Blob*, 1958

On a clear spring day, I make my way to the Hunterian Museum in Glasgow. It's located in a part of the university campus which has a whiff of Hogwarts about it, with a maze of courtyards, the entrance hiding somewhere within. The Hunterian is the oldest public museum in Scotland, more cabinet of curiosities than modern scientific institution. Roman artefacts, feathered Maori cloaks and fossils sit side by side in this museum, which reminds me of a church with its pointy gable and simple rosette glass window, the high ceiling worked in the same dark wood as the carved balustrades. But it's neither the neo-Gothic charm nor the wonderful collections that brought me to the Hunterian. I am here to see a glass bottle, about as big as a hand, fitted with a fat stopper and two gilded labels, handwritten. I'm here to see an old phial filled with slime.

Some forms of matter seem to unite the properties of solids and liquids. For example, cats: which category do they fall into? The answer should be an easy one, physically speaking: solids retain their shape, while liquids fill their container. Cats seem unequivocally to be solids, until they demonstrate their aptitude for easily slipping into the smallest of gaps, almost flowing into them. The French physicist Marc-Antoine Fardin researched, tongue-in-cheek, the physical classification of cats between solid and liquid, touching on his specialist area of rheology, the study of the flow of matter. In 2017 he was awarded the Ig Nobel Prize, a not altogether serious award for suitably original research.

Matter which exists somewhere between solid and liquid

is not restricted to cats, though, and slime is its most important embodiment in nature. It is protean in its behaviour; it is the material of interfaces and has a unique place in our imaginations. We are all creatures of slime, but some of us are more creative than others: there is a menagerie of oozing organisms to be found in all the world's habitats, frequently changing these environments to suit their needs – by leaving their glistening marks. It may also be a surprise to discover that microbes were not only the dominant, but also the only form of life on Earth for billions of years, with slime, as the *éminence gluante*, propping up their power and setting in motion processes across the globe which still shape it today.

The long reign of slime concerns the supposedly boring stages of evolution which preceded the emergence of the first animals. Popular portrayals often neglect this seemingly endless span of time, but slime was paving the way for life on Earth then, particularly for higher organisms like us. It may even have facilitated our very existence. It is a legacy we humans prefer to ignore. Here, we benefit from slime's hidden nature, with its visible manifestations banished inside our bodies. Better not to think about slime, which seems to carry the scent of our baser instincts, of sex and weakness, of sickness and death. We regard it as an object of universal disgust, only admitting it – at least in industrialized societies – into our hyper-hygienic worlds in controlled doses, as an entertaining expression of the depths of the human psyche.

Modern monsters can rarely do without slime and slobber, be it on screen in movies like *Alien* or in stories like the ones H.P. Lovecraft wrote. In a sense, slime makes humans biological creatures, yet becomes the line of demarcation between us and the Other. Is this because slime as a phenomenon is slippery to grasp but nonetheless elicits strong emotions? Physically speaking, it can be defined and therefore contained. Slime is an extremely aqueous and viscously fluid hydrogel, which also bears the properties of a solid under certain conditions. Biological slimes are so flexible that they can easily adapt as required. Scientists are trying to copy or emulate these sophisticated structures for applications like soft robots, smart wound dressings or tailor-made glues, but often come unstuck when attempting to unravel their biological complexity.

Yet the first steps have been taken and, in the future, it may be possible to imitate even more specialized slimes, which serve as the adhesives, lubricants and selective barriers vital to microbes, animals and plants. There is probably no single living creature that does not depend on slime in some way. Most organisms use slime for a number of functions, be it as a structural material, as jellyfish do; for propagation, as plants do; to catch prey, as frogs do; for defence, like the hagfish; or for movement, like snails. The ubiquity of slime is little recognized, because many biological hydrogels hide behind pseudonyms like 'mucilage', 'mesoglea', 'marine snow' and, of course, 'mucus', which barely give an inkling of their true and common nature.

Since many of these biological hydrogels are secreted outwards, they work beyond the single organism. Even in the natural environment they are invisible cement, holding different ecosystems together, from desert to coastline to marine habitats, primarily at the interfaces where water, land and air meet. Slime is a central cog in the world we live in and even slight changes could have global effects. The looming reality of climate change and other environmental crises like the loss of ecosystems and biodiversity threatens hydrogel-based relationships and processes. However, a new equilibrium in a warmer world might also favour slime in some habitats, allowing it to return to dominance. It would be a step back into an early era of evolution, a new *era of slime*.

The importance of slime is similarly far-reaching inside the human organism, equipped as it is with four different gel-like systems to fortify it with living walls, gates, armour and moats. Most pathogens founder at these defences, while welcome microbes find refuge as tenants and mercenaries. Like materials science and climate research, biomedicine has been waking up to its own 'slime blindness'; defective hydrogel barriers have an important part to play in infections, chronic intestinal inflammations, cystic fibrosis, cancer, heart attacks and a series of other ailments.

Without slime we would be overrun by pathogens: hydrogels are sticky traps for microbial invaders which actually produce slime as well. Keeping our distance from unknown hydrogels is expedient and our aversion to all things slimy well-founded, as the science of disgust shows. But does even the word 'slime' have to elicit gagging histrionics? Our all-too-often highly emotional response to viscous materials can lead to ignorance as we forget that slime is the very foundation of our biology, an essential for our health and our environment. And that kind of ignorance is a very modern idea.

In Ancient Egypt, mud and slime were thought of as a source of life. Aristotle's works on the theory of spontaneous generation of insects and other lowly animals was then supported for many centuries. It received an unexpected boost when Charles Darwin's theory of evolution overthrew biblical ideas on the origins of nature. Could primordial slime on the ocean floor be the source of life? This claim was made by the prominent Darwinist Ernst Haeckel (1834–1919) and it electrified early evolutionary biology for a brief period in the second half of the nineteenth century.

Haeckel was fortunate to live in an age when the technological revolution opened access to the deep sea for the first time. Samples from the floor of the Atlantic Ocean seemed to confirm his theory. They contained a gelatinous mass which he quickly declared to be primordial slime; it even seemed to move – especially when observed on board a ship on the high seas. It was Haeckel's misfortune to live during the dawn of the discipline of oceanography and its rigorous analytical technology. His theory was tested when HMS *Challenger* embarked upon an expedition lasting many years and spanning the globe. Instead of charting the habitat of primordial slimes, it ultimately exposed Haeckel's hypothesis as a boozy mistake.

★★★

In Glasgow around 145 years later, it is this specimen that I'm hoping to trace: the find which disproved the theory of primordial slime as giver of life. On 4 March 1876, the crew of HMS Challenger *brought up some matter from the ocean floor of the South Atlantic. To their surprise, the specimen did not contain any pulsating primordial slime; it was just seawater. Only once the sample had been properly preserved in plenty of alcohol did a gelatinous mass appear inside the glass vessel. But it was merely a mineral precipitate, resulting from a reaction between seawater and alcohol. And so the brief era of primordial slime was over. The only things left are an amusing anecdote and the specimen from the* Challenger *expedition, which I struggle at first to find at the Hunterian, though I comb through every cabinet in every hall and in the gallery.*

The primordial slime phantasm wasn't Haeckel's sole contribution to early evolutionary biology – nor the most controversial, either. He was one of the most prominent advocates of an 'evolutionary racism' that accorded human lives unequal value. Nowadays he is best known as a natural scientist and gifted artist who made gelatinous sea creatures so popular in science and among the public that art nouveau took inspiration from his drawings of their graceful silhouettes. Haeckel based his system of nature on primitive organisms, little more than the smallest units of slimy protoplasm. This was the name given to the gelatinous, pulsing substance inside cells, which according to a popular theory was the most basic expression of life.

As such, protoplasm was even believed to be capable of receiving invisible signals from the outside world. Towards the end of the nineteenth century, a new and exciting age dawned; people began to think of themselves as surrounded and permeated by vibrations and oscillations coming from the ether, an invisible and possibly gelatinous medium thought to span the entirety of space. If the transmission of sound, light and energy could be traced back to oscillations, would similar vibrations influence the protoplasm, the substance of life itself? According to some theories, their traces

– a bit like grooves on a vinyl record – created individual patterns of experience, to be stored and bequeathed to the next generation.

When it came to unseen impulses from the ether, artists were considered to be extremely sensitive and even capable of perceiving the world as a whole. Luminaries like the painter Edvard Munch tried to find a new science-based language for the arts to convey that experience to their audiences. Psychic mediums were another conduit for the ethereal realm, all the more believable when a kind of protoplasm flowed plentifully from their bodies. Or so it seemed. The First World War brought a boom in contacting the other side, while protoplasmic collaborations between empirical sciences and the arts came to an end. The world of trenches and death marches through muddy 'slimescapes' was one of men hard in body and spirit, without a place for seemingly feminine softness and yielding.

Twenty years ago, an article on snails in the *New Scientist* prompted me to think about slime not as some unknowable phenomenon but as a material with specific chemical compounds, physical properties and biological functions. I thought about writing a book then, but envisioned some sort of glorified leaflet or small booklet at most. What was there to know about slime, anyway? But then ever more aspects of this fascinating material revealed themselves to me, often as chance discoveries; one detail would lead me to a new anecdote, which was connected to a little-known publication that opened yet more doors.

Slime and evolution: spanning billions of years. Slime and the planet: glueing global cycles and processes together. Slime and life: a foundation to us and all organisms. Slime in the lab: technology going soft. Slime on paper: linking nature and art. Slime and monstrosity: a trigger of disgust. Little by little, enough pieces of the puzzle came together: slime is neither an accident nor an exceptional presence in the world. It is an omnipresent rule. That makes a definitive book on slime an improbable feat. The material is too varied, its evolution too long and our love-hate relationship with it too volatile.

This book is an attempt to span more than one great arc with chapters that can be read in sequence or independently. Not every reader will be interested in every single facet of slime and might skip some parts. This is why I repeat fundamental terms and processes as required. But reader, be warned: this is a sprawling story, bursting at the seams and hard to contain at times. When it comes to slime, there are no hard borders or distinct divisions, or a chronology to connect the dots. Please indulge yourself with the dripping and oozing glory of slime. You won't get lost if you follow the signs.

The *Phenomenon* section explores slime as an often emotionally charged concept and monstrous prop. *Physics* examines the material and its unique properties, while *Organism* looks at the significance of gels for the body. Before disgust came high regard, as the origins and the very essence of *Life* were traced back to primordial forms of slime, in science and modern art. Slime is indeed a companion of all life, a pioneer of *Evolution*, whose secrets we have only begun to unlock. So, has *Nature* spawned a single slime-free creature? It's hard to imagine, because gels have many uses, depending on habitat and organism. They also shape the outward appearance of our *Environment*. Climate change and other catastrophes threaten this balance, but could also work in slime's favour, ushering in a new era of gooey dominance.

★★★

Primordial slime? Never heard of it, replies a staff member at the Hunterian when I enlist her help. But she begins to search, even recruiting two colleagues, themselves equally in the dark when it comes to the slime's location. We fan out and search for a long time before they finally bring me the good news that there are several bottles of historical goo. Could my slime be among them? And they're right: I find the phial from the Challenger *expedition in a glass cabinet in the main hall. It's so high up and out of the way that I had missed it altogether. It is surrounded by other historical treasures from marine research trips: dried mud from the ocean floor and a pickled lungfish, dolefully observing the goings-on below. One of the staff gets me a ladder, bringing me eye to eye with the slime. In storage for so long, the gelatinous mass has disintegrated, and the bottle looks to be filled with clear water. Only at the bottom is a delicate, shimmering layer of white, barely perceptible to the eye. It seems fitting that this historical goo no longer exists, since primordial slime never really did either. It's also fitting for many of the other slimes in this book, which, at least when I embarked upon my research, I could sense more than see. Slime rarely has its place in the spotlight, though it holds us and our worlds together. It's high time we run towards the blob and take a closer look.*

I
Phenomenon

Slime is a biological substance with physical properties and biological functions. But it is also a phenomenon or an idea which repels and excludes. This has made it an object of disgust in popular culture, be it in literature, in comics or in film. In recent decades, monsters of all kinds can be found leaving extraordinarily slippery trails of slime. Slime has entertainment value, but in playing with our aversions it also offers a kind of lusty frisson. Disgust, one of the most basic of emotions, is intended to protect us from pathogens and infections, but we feel it in response to much more than potential microbial contaminants. Disgust at the crossing of social boundaries can spark discrimination. At worst, people and even whole groups are stigmatized. A long and inglorious tradition has seen women forced into this category – as the apparently slimier sex.

1

Cosmic Horror

'He slimed me.'
'That's great! Actual physical contact!'
From *Ghostbusters*

An abandoned corridor in a New York hotel plays host to a close encounter of the more unsavoury kind when ghostbuster Peter Venkman, still a rookie on the job, finds himself knocked down by a Class-5 Full Roaming Vapor and covered in green goo. This slimy showdown from supernatural comedy *Ghostbusters*, released in 1984, emerged at a time when almost every film with a hint of spookiness was swimming in gunk. Cinemas everywhere unleashed a great tide of slime, swamping a happily revolted public. The American journalist Daniel Engber views this tsunami of slime as an expression of a volatile era, one in which humanity's survival seemed under threat from radioactive contagions.

These fears found numerous manifestations in popular culture, sometimes, according to Engber, as luminous slime. Perhaps the *Ghostbusters* goo was vague enough to serve as a surface on to which diffuse emotions of all kinds might be projected, from a fear of radiation to a perennial aversion to slime itself. Yet the backdrop of history isn't the only reason why Peter Venkman's encounter with the slimy Class-5 Entity is so effective: even if you watch the cult classic today, it is nauseating. There's something about slime that keeps us glued to the screen. It seems to be an unwritten rule of horror: slime sells! Zombies ooze, while mutants, aliens and other beasts are made of goo or leave, at least, a trail of slobber to get their audience gagging. As Jane Austen didn't quite say, it is a truth universally acknowledged that a monster with a nice leading role must be in want of some slime.

The toy industry understands this perverse allure too. It has well and truly put its money on the goo, the slimier and more disgusting the better. It may be little more than rubbish in the eyes of adults, but that is precisely its appeal for children; it's about creating a distinction between their world and the world of adults. In recent years, whole generations of children, from primary-school age to teenagers, have boosted the production of shimmering, glowing or outright toxic-looking slimes of all kinds. The thrill is in the taboo, in crossing a line, in grabbing it with both hands.

It's similar on the page, as the horror writer Stephen King freely admits. After terror and horror, the 'gag reflex of revulsion', he writes, is the final facet of the goose-bumpy genre which he inflicts on his audience: 'I recognize terror as the finest emotion and so I will try to terrorize the reader. But if I find that I cannot terrify, I will try to horrify, and if I find that I cannot horrify, I'll go for the gross-out. I'm not proud.'

Of course, Stephen King didn't invent the doctrine of disgust. That occurred long before he took up the pen or green ectoplasms slunk on to our screens. The creations of H.P. Lovecraft (1890–1937) set the standard for slime in literature. The American author achieved little success during his lifetime, only publishing his stories in cheap pulp magazines. It's something of a tradition: slime continues to make an appearance in the apparently lesser fields of Gothic fiction, as well as in children's books, comics, computer games and pop culture of every stripe. Lovecraft is now regarded alongside authors like Edgar Allan Poe as one of the forefathers of fantastical horror-writing. Stephen King himself was inspired by 'the twentieth-century horror story's dark and baroque prince'.

Lovecraft's writing is populated by monstrosities spawned from an especially dark imagination, like the amorphous shoggoths in *At the Mountains of Madness*. These creatures are not much more than deep-black slime, out of which limbs can be formed: 'a shapeless congeries of protoplasmic bubbles, faintly self-luminous, and with myriads of temporary eyes forming and unforming as pustules of green light all over the tunnel-filling front that bore down upon us . . .'. Storylines are woven through Lovecraft's novellas and short stories which centre on the legend of Cthulhu, a kind of winged deity with tentacles on its face.

Cthulhu itself sleeps in his lair, the 'green, slimy vaults' in the underwater 'corpse-city' of R'lyeh, dreaming of his return – when the stars align. *The Call of Cthulhu* gives the reader a taste of this, when the powerful creature is torn from its slumber by a passing ship's crew: 'Everyone listened, and everyone was listening still when It lumbered slobberingly into sight and gropingly squeezed Its gelatinous green immensity through the black doorway into the tainted outside air of the poison city of madness.'

'Lovecraftian' has become a synonym for cosmic horror, a genre which sees ancient creatures from outer space or obscure gods living among us. They attempt to take over the world, abetted by wilful ignorance when humans turn a blind eye, allowing evil to take root and thrive in our midst. Lovecraft continued to reformulate this tried-and-tested horror motif time and again and inspired many of the more recent classics of the genre. His influence is particularly striking on the big screen, where our planet is overrun again and again by creatures from the depths of outer space, that is to say, from the depths of our subconscious: these are our fears made manifest. And at the centre of it all is slime.

The ectoplasmic escapades in *Ghostbusters* are triggered by the return of the Sumerian god Gozer to Earth. He finds a temporary home in the form of Sigourney Weaver, who would soon go on to encounter yet more drool in another iconic scene in *Alien*. But slime's first big break on screen was as the deep-red *Blob* from outer space, which crash-landed on Earth in 1958 and proceeded to hunt a young Steve McQueen and other earthlings. *Run, don't walk*, the trailer warned, decrying the 'red menace', presumably symbolizing communism. While the ghostbusters were slimed by ectoplasmic fears of nuclear waste, modern killer slimes, as conceived by the American writer Jeff VanderMeer, spring from biotech labs or Mother Nature's overheated mutation engine. Slimes go with the times, offering up a blob for every era and its attendant paranoias.

Stanisław Lem's philosophical novel *Solaris* is another great work of science fiction which has escaped any explicit interpretation. In the story it is humans that infect an extraterrestrial planet, ruled by an all-encompassing ocean of gel, which creates structures of mysterious beauty kilometres high. For decades, every attempt to communicate with this strange and totally indifferent intelligence fails. The fictional mission does, however, achieve some, if painful, success. According to the famous astrophysicist and author Carl Sagan, ventures into the cosmos always hit rather close to home: 'The quest for extraterrestrial intelligence is, in essence, a quest for ourselves.' If so, then Lem testifies to human limitations.

Our own weakness is also the subject of H.P. Lovecraft's novella *The Shadow over Innsmouth*, with the protoplasmic shoggoths in a cameo role. They serve amphibious fish-people who live in the depths off Devil's Reef near an isolated little town in New England, passing on a toxic legacy to its inhabitants. Escaping the curse is impossible, as the hero of this nightmarish story is ultimately forced to accept. It is probably a fictional echo of the author's real-life fears. Born in Providence, Rhode Island, Lovecraft's childhood and youth was overshadowed by his father's long illness and early death. The elder Lovecraft was taken to an insane asylum and is believed to have died there of the effects of late-stage syphilis.

Men who suffer from the disease are able to infect their wives, who might in turn pass the pathogen on to their unborn child. Did Lovecraft's mother Sarah Susan search for signs of her husband's 'degeneracy' in her son? All we know is that she swung between showering him with love and describing him as 'grotesque' and 'disfigured'. It is perhaps no surprise that H.P. Lovecraft developed into a night-walking loner. His mother died, like her husband, at Butler Hospital in Providence, though it remains unclear as to whether she too had contracted syphilis.

The couple's son, however, had found solace in books from an early age. Alongside the horror classics of Edgar Allan Poe were articles on astronomy which may well have provided Lovecraft with the inspiration for outer space as a source of all-powerful monsters. Their biologistic, slimy appearance might be traced back to an awareness of evolutionary biology, which was becoming increasingly established among the public at the time.

Ernst Haeckel, the leading German Darwinist, was especially influential. Many of the undulating lines and exquisite details of art nouveau designs of the time are, as we have seen, indebted to his biological drawings. But his accomplishments as an illustrator do not permit us to forget his work on the origin of man, which he based on breeding and selection and led him to categorize humans into 'species' and 'races', with some inferior to others and even doomed to extinction. He also seemed to suggest that disabled people were an unsupportable burden on society. Whether he carried a thought experiment through to its bitter climax or was writing with conviction remains unclear and is ultimately irrelevant. His work was among that which formed the theoretical basis of twentieth-century eugenics.

Haeckel found a keen student in Lovecraft, who romanticized himself as the noble descendant of the New England aristocracy, born in the wrong period of history. Lovecraft rejected modernity. The then-dawning American Century and the First World War appear as little more than background noise in his fiction, which he historicized in almost comically stilted language. What perhaps began as eccentric retrophilia on Lovecraft's part, however, had a much darker side as well. Racist and anti-Semitic, he was repelled by a mix of ethnicities. That which was foreign could not be assimilated in his view. There is an often-quoted letter from the 1920s about the inhabitants of the New York slums whom, in lines dripping with poison and disgust, he calls a 'nightmare of perverse infection':

> The organic things – Italo-Semitico-Mongoloid – inhabiting that awful cesspool could not by any stretch of the imagination be call'd human. They were monstrous and nebulous adumbrations of the pithecanthropoid and amoebal; vaguely moulded from some stinking viscous slime of earth's corruption, and slithering and oozing in and on the filthy streets or in and out of windows and doorways in a fashion suggestive of nothing but infesting worms or deep-sea unnamabilities. They – or the degenerate gelatinous fermentation of which they were composed – seem'd to ooze, seep and trickle thro' the gaping cracks in the horrible houses . . .

In Lovecraft's fiction, as in other horror stories, it is usually hard men, most often scientists, who come face to face with cosmic killer slimes. Their boundless curiosity and hubris are what uncovers nature's dark secrets in the first place and unleashes horrors which should have been kept under lock and key – or eternal ice. In 1951, in the eponymous movie, Norwegian polar scientists stumbled on screen upon the Thing from Another World amid the unending Antarctic ice, where it was hibernating in its spaceship. Like the shoggoths, once awakened this alien can shape-shift, impersonating the men themselves with beguiling accuracy. From then on, the scientists must fight a monster which walks among them, assuming a familiar face.

The drippy evil residing inside us gets even more personal in the David Cronenberg film *The Fly* from 1986. Here, scientific hubris brings about the downfall of Seth Brundle, a scientist, who transforms into a grotesque, oozing human-insect hybrid after an experiment gone awry. Like the inhabitants of Lovecraft's Innsmouth, Brundle metamorphoses, changing from a larval-stage human into a degenerate monster. *The Fly* was a success, but also marked the end of the cinematic orgy of ooze. The tide of slime which had washed over the big screen seemed to retreat as abruptly as it had come.

'The end of the millennium heralded the closing of a cycle,' writes Rebecca Bell-Metereau, a media studies scholar, in an essay on slime, sexuality and the grotesque in 2004, continuing: 'Audiences seemed to lose their fascination with gross and slimy fare. Apocalyptic fears and terrorist attacks dampened viewers' desires for images of the grotesque. Supplanted by a crop of war pictures, these slimy, disgusting movies passed out of fashion for a time . . . but they will be back.' From today's perspective, Bell-Metereau's assessment has proved exactly right.

Alien never really went away, new versions of *The Blob* and *The Fly* are planned, and *Ghostbusters* recently returned to screens with an all-women team hunting down slimy Class-5 Entities. Even some more of Lovecraft's hard-to-adapt stories are supposedly in the works to appear on screen. And slime goes small nowadays by streaming right into our homes. *The Expanse* on Amazon is a series set in a future that has all but forgotten timeless Lovecraftian lessons. Human factions fight each other over a luminous gel that turns out to be a highly infectious alien slime with a mind of its own: 'Take that, blue goo!' On Netflix, on the other hand, the inhabitants of small-town Hawkins, Indiana equally don't get much choice but to fight the black slime oozing its way in from a scary 'Upside Down' world – *Stranger Things* indeed.

★★★

It is a dark and dystopian mirror to our own world that slime represents here, just like it often embodies the blurred line between us and the Other. It is, in fact, a role it also performs in the natural world. After all, we are biological beings of an amorphous nature, which neither the soft shells of our bodies nor the corset of culture can permanently contain. 'Civilisation, control and safety form a crust over deep waters,' writes the author Nicci Gerrard. 'In all of us, often pushed into the corners of our minds, is the uneasy awareness of how frail our hold over ourselves is, how precarious a grip we have on our own minds and bodies.'

Our boundaries are fluid enough in reality. And they are marked in slime. During sex, slime is the boundary between me and you, even desirable in the moment for frictionless contact. Slime often highlights the transition from good health to sickness. And it represents the crossing of the last barrier between life and death, when the body loses all definition during slimy decomposition. There is one decisive difference, though: while living organisms produce and secrete extremely varied and complex gels, the slime of death and decay lacks any sophistication. It represents biological breakdown and a loss of order. It is a waste product.

It is not only in the body but also in the mind that slime has to embody weakness. It stands for supposedly primitive and animalistic impulses that compete with our stronger and nobler disposition. It becomes a question about the distinction between the 'I' that we are, and the 'I' that we would like to be – or lack thereof. Dr Jekyll, the protagonist in Robert Louis Stevenson's 1886 novella, is forced to learn that painful lesson. In a fit of scientific hubris, he attempts to separate the good in his soul from the evil. Instead, he helps bring about the murderous Mr Hyde, who not so much emerges 'out of the slime of the pit' as takes visible form for the first time. The monster has never not been there. He was simply shackled deep inside good Dr Jekyll.

The French philosopher Jean-Paul Sartre was preoccupied with the notion of slime, raving for several pages of his 1943 essay *Being and Nothingness* about 'le visqueux'. It is the viscous, fluid stickiness – that is to say, slime – which caused Sartre so much existential angst. Slime, 'the agony of water'. Slime, the 'degenerated liquid'. Slime, 'like a leech sucking me'. According to Sartre, 'the slimy flees with a heavy flight, which has the same relation to water as the unwieldy earthbound flight of the chicken to that of the hawk'. He goes further, describing slime as 'a soft, yielding action, a moist and feminine sucking', or even 'a sickly-sweet, feminine revenge'.

Is this a philosopher's cool-headed view of one phenomenon among others and its qualities, or was Sartre's revulsion so deeply entrenched that it foiled any attempt at sober observation of 'le visqueux'? When the philosopher speaks of the fear of losing oneself or even disintegrating in stickiness, is he talking about himself and his own Monsieur Hyde? When he mentions being 'slyly held back by the invisible suction of the past', is he haunted by his own sins? And the Blob once more strikes back.

* * *

But is it always slime which maketh the monster? The great Czech writer Karel Čapek was born in 1890, the same year as H.P. Lovecraft, to whom he even bore something of a resemblance with deep, shadowy eyes set in a serious face beneath a head of neatly slicked-back hair. In 1936, both authors published significant works boasting surprising parallels. Where Lovecraft wrote of amphibious beasts living in Innsmouth near Devil's Reef on the New England coast, Čapek's *War with the Newts* has his monsters inhabiting Devil Bay in Sumatra. Yet Čapek's slimy creature is based on the anatomy of a real animal, the extinct *Andrias scheuchzeri*. The giant salamander's fossilized remains had been discovered in the eighteenth century and were, for a time, interpreted as the bones of small people.

In the novel, however, the creatures hope to wage war on humanity and strive towards world domination. So far, so Lovecraft. Yet unlike Lovecraft's monsters, Čapek's newts are peaceable creatures, until they are exploited, enslaved and abused by man. 'Then it can be cooked or stewed, and tastes like inferior beef. In this way, we ate a Newt called Hans,' a scientist explains in the book. 'He was an able and intelligent animal with a special bent for scientific work . . . We used to have long conversations with him in the evenings, amusing ourselves with his insatiable thirst for knowledge. With deep regret we had to put Hans to death, because my experiments on trepanning him made him blind.'

Who is corrupting whom? Who is the real monster here? Both writers were witnesses to at least one world war. But, unlike Lovecraft far away in America, with slaughter and catastrophe re-emerging in Europe Karel Čapek did not need to conjure up alien intruders from other worlds. Fantastical as his stories are, they are timeless tales and warnings about man becoming man's deadliest enemy again and again. Čapek warned against fascism and human cruelty in his fiction: 'Two dead salamanders were left on the shore and one with a broken spine, which emitted a sound like "ogod, ogod, ogod".' Čapek was regarded by the Gestapo as one of the most dangerous men in the country. When he died of pneumonia in December 1938, a warrant had already been issued for his arrest.

2

Dr Jekyll and Mrs Slime

Hortense and Edgar were making love, Edgar reaching down from a little rock to kiss Hortense on the mouth. Hortense was reared on the end of her foot, swaying a little under his caress like a slow dancer enchanted by music.

Patricia Highsmith, *Deep Water*

Patricia Highsmith's great love of snails began as a young woman, when she came upon two brown-striped specimens locked in 'a bizarre embrace' at a New York fish market and took them home. Her fascination for the creatures' mating ritual inspired two of her protagonists. The first was Victor Van Allen in *Deep Water* (1957), who observed an intimacy in the animals which was lacking in his destructive marriage. In her short story 'The Snail Watcher', however, the ominous beginning hints where the short story is going: 'When Mr Peter Knoppert began to make a hobby of snail-watching, he had no idea that his handful of specimens would become hundreds in no time.' Spoiler alert: they don't stop till they overwhelm their host in a 'glutinous river'.

Highsmith herself was unafraid of slimy armies, keeping hundreds of snails as pets and taking them in a salad-filled handbag to dinner parties for company. When she moved to France and wasn't allowed to bring foreign *escargots* legally into the country, she even resorted to smuggling. In several trips she brought the snails across the border, stowing a few of them under her bosom each time. A surprising degree of care, perhaps, for a woman with an abrasive personality, but maybe less surprising for an author who felt drawn to the grotesque, the gruesome and the macabre.

According to Fiona Peters, Professor of Crime Fiction at Bath Spa University, the snail represents an archetypical grotesque creature in Highsmith's fiction, keeping the audience between laughter and disgust or fear. But there is also a long tradition of the snail as a symbol of transgressive feminine sexuality. This would make the animal 'abject' like other objects that – as the French-Bulgarian philosopher Julia Kristeva describes it – disgust or horrify because they blur the distinction between Self and Other. In a sense, it could have been a similar question of attraction and repulsion when Sartre compared the viscous to the 'flattening-out of the over-ripe breasts of a woman lying on her back'.

Highsmith was deeply influenced by Sartre's writings, but it is to be hoped that he never came across her work like the short story 'The Quest for Blank Claveringi', where an over-ambitious scientist seeks out carnivorous giant snails on an otherwise deserted island. There are valuable insights for him to be gained even though he never gets a chance to report them back: the animals do exist. They are indeed carnivorous. And they are surprisingly fast, especially if the snail in question is a female avenging her murdered partner. 'The wonderfully macabre ending neatly upends the human propensity for eating snails,' Peters writes. 'Just as snails are starved, boiled alive and then swallowed, the man here gets drowned and chewed to death by thousands of teeth at the same time.'

***

What we all seem to fear is the animalistic parts of our bodies and minds and the slimy ways in which they show themselves when we are at our weakest. But resistance seems futile, as the philosopher Martha Nussbaum writes in *The Monarchy of Fear*. We are deceiving ourselves when we think of ourselves as above the animals. Every day in countless ways our animal bodies compel our attention, as we make almost constant contact with our own secretions and those of others. But the need for distinction is forging ahead with catastrophic consequences. 'Human beings are probably hard-wired to find signs of their own mortality and animality disgusting, and to shrink from contamination by bodily fluids and blood,' Nussbaum opines in the *Washington Post*. 'But in every culture something worse kicks in: the projection of these feared and loathed characteristics on to a vulnerable group from whom the dominant group wishes to distance itself.'

This kind of stigma is forced on powerless minorities, manifests itself 'in racism, in homophobia and even in revulsion toward bodies of people who are aging', as Nussbaum writes. 'But in every culture male disgust targets women, as emblems of bodily nature, symbolic animals by contrast to males.' She sees disgust for women's bodily fluids as fully compatible with sexual desire, when 'men often crow with pride over intercourse with a woman imagined as sluttish and at the same time defame and marginalize her'.

This is a contradiction even Jonathan Swift caught in his poem 'The Lady's Dressing Room', written in 1732:

> Should I the Queen of Love refuse
> Because she rose from stinking ooze?

An amusing inconsistency, but also an ostracism with a long, Christian history. Around 900 CE, the Benedictine abbot Odo of Cluny is said to have argued that 'The beauty of the body resides wholly in the skin,' continuing:

> In truth, if men could see what is beneath the skin, the very sight of women would be nauseating to them. Their grace consists of phlegm and blood, wetness and gall. Consider what is hidden in the nostrils, the throat, the belly: filth throughout. And we, who loathe to touch vomit or manure even with a fingertip, how could we desire to clasp a very sack of excrement in our arms?

The basis for this somewhat idiosyncratic analysis are the teachings on the four humours, which were established in antiquity and endured for centuries. The teachings stated that the human body was made from four humours on whose balance our health depends: yellow and black bile, blood and phlegm – which is slime.

The slimy humour was and is of course common to both men and women, but even in the nineteenth and twentieth centuries men like Sartre and Lovecraft associated it primarily with the other sex. Known for his romantic trysts, the French philosopher once said he preferred masturbating women to the act of sex itself. In her book *At the Existentialist Café*, Sarah Bakewell writes: 'Sartre, if we can judge by the vivid descriptions in his books, found sex a nightmarish process of struggling not to drown in slime and gloop.' A similar fear might also have plagued H.P. Lovecraft, who considered himself a latter-day Puritan.

In his short story 'Dagon', a sailor is stranded on an island covered in black slime and succeeds in escaping from a monstrous, oozing creature at the last moment. Shortly after his return to humanity, the lonely hero discovers the legend of the Dagon, the fish-god, which proceeds to haunt his nightmares. The very thought of the deep ocean makes him shiver 'at the nameless things that may at this very moment be crawling and floundering on its slimy bed'. The story ends with an ominous noise, 'as of some immense slippery body' pushing against the door of the room. So who is forcing their way in? A fish-god, a woman or a slimy symbiosis of the two? Could it be that Lovecraft was referring to the mythical Dagon, a Mesopotamian deity, always depicted as a hybrid creature, sometimes a cross between a fish and a woman?

This monster presented Lovecraft with a perfect blend of all his most deeply personal fears and aversions. There's the sea: the young Lovecraft is reported to have told a friend that he hated fish and the sea and feared anything to do with them. And there's physical closeness: on one occasion, Lovecraft described adulthood as hell, though he did then go on, entirely inexperienced, to marry hastily in his mid-thirties. The relationship was probably doomed to fail, even though the writer lived up to his name by reading up on the art and craft of love in books. Even so, after their divorce his ex-wife Sonia Greene bemoaned his lack of interest and initiative.

Lovecraft's biographer, Sunand T. Joshi, has gone so far as to describe Lovecraft as one of the most asexual individuals in human history. Neither women nor sex play much of a role in the horror writer's works. But what can be read between the lines? '[S]ex that is sometimes presented in disguised, Freudian terms, such as Lovecraft's vaginal creation, Great Cthulhu,' writes Stephen King, making new enemies among Lovecraft's ardent fanbase. 'After viewing this many-tentacled, slimy, gelid creature through Lovecraft's eyes, do we need to wonder why Lovecraft manifested "little interest" in sex?'

It seems clear that both Sartre and Lovecraft exhibited an at best ambivalent relationship to the female body and nursed deep-seated aversions to sea creatures. Sartre, when asked in an interview by his lifelong partner, the French writer and philosopher Simone de Beauvoir (1908–86), what nauseated him – other than tomatoes – responded merely, 'crustaceans, oysters, shellfish'. In fact, he was terrified of marine life from an early age and triggered that phobia much later with a mescaline injection. That particularly bad trip lasted for days, making him hallucinate about swarms of giant crabs. And a demented octopus. But there is more. When asked by de Beauvoir if his disgust of shellfish had anything to do with what he thought about mucus and viscidity, he agreed: 'Certainly.'

This can't and shouldn't be more than speculation: but could slime have bridged the gap between the deep-seated aversion to sea life and sexual reserve? For Sartre, slime was a fear-laden phenomenon which he eschewed in sex. For Lovecraft, it must have seemed repellent enough to use repeatedly as a feature of his slippery monsters, which often come from the sea. Some foods are thought to be reminiscent of human genitalia, and therefore stimulating. The glistening, fleshy oyster supposedly resembles the female sex and oozes slimy fluids. It has long been considered an aphrodisiac, in both Lovecraft's America and Sartre's France.

It is France where, incidentally, another firm fruit with slimy innards is known as the libido-boosting apple of love or *pomme d'amour*: the tomato. It does seem plausible that H.P. Lovecraft and Jean-Paul Sartre were as repelled by viscous fluids on their plates as in their beds. Yet finding 'the second sex' repulsively slimy, with allusions to marine life, is not the preserve of men: 'Feminine heat is the flaccid palpitation of a shellfish . . . she is sucking, suction, sniffer, she is pitch and glue, immobile appeal, insinuating and viscous', wrote Simone de Beauvoir.

And the men? Long before she was famed for her thrillers with slick psychopaths and relentless snails, the sixteen-year-old Patricia Highsmith submitted to her first kiss. According to her biographer, Andrew Wilson, Highsmith's mother had recruited the young man as her daughter's date for the evening, since she was showing disappointingly little interest in men. The teenager played along, but on her return described the kiss as 'like falling into a bucket of oysters'. It was too high a price to pay for being invited to dinner. She would rather pay for herself.

Still, she wanted so badly to prefer men: Highsmith wrangled with her sexuality her whole life, even trying to therapize away her desire for women. Her numerous affairs and relationships with partners of both sexes were mostly short-lived and often left a bitter taste because Highsmith took pleasure in forcing couples apart. She died of cancer, alone in a clinic in Locarno, Switzerland, her difficult character no doubt contributing to her isolation at the end.

She preferred to write her stories and books in bed – almost in the foetal position, according to Highsmith herself – with ashtrays and hard drink close to hand. The bed, she wrote, was 'a womb of my own'; not only onomatopoetically coming close to the 'room of one's own' that Virginia Woolf declared to be crucial for a woman to write, alongside a stable income. Highsmith may have felt secure in her bed, dreaming up sly and slimy murderers. She called at least one of her novels a 'nine-month-old foetus in the womb of literary creation', Wilson reports. But she didn't have children, whom she found abhorrent in real life, and had them appear in her stories as little more than extras.

The paintings of the Mexican painter Frida Kahlo are quite different in this regard, since many of them feature children that are painfully absent. Her work centres on the bloody emptiness, the heavy suffering that numerous miscarriages left behind. Female reproduction is a great omission in the canon of Western art, if bracketed by mostly pornographic sex and often mythically glorified motherhood. Where is the bloody and slimy reality of women's bodies, be it menstruation or birth? Its absence reflects the ambivalent relationship we have with bodily, or at least, female fluids. 'The shedding of blood has historically been seen as a male act of heroism: from right-of-passage fistfights to contact sports and combat,' writes the author Sinéad Gleeson. 'Female bleeding is more mundane, more frequent, more get-on-with-it, despite its existence being the reason that every single life begins.'

Not too subtly, Charlotte Roche's bestseller *Wetlands* turns this around and takes it to an extreme. In the book, female slime, blood and other bodily secretions are not so much celebrated as illuminated in lurid colour. There is just one notable exception: the novel's protagonist is quick to report that she had herself sterilized the day she turned eighteen. Here, the renunciation of birth and motherhood is voluntary, anticipating another tabooed phase of female reproduction – its end. 'The menopause is probably the least glamorous topic imaginable,' writes author Ursula K. Le Guin (1929–2018):

> The woman who is willing to make the change must become pregnant with herself, at last. She must bear herself, her third self,

her old age, with travail and alone. Not many will help her with that birth . . . That pregnancy is long, that labour is hard. Only one is harder, and that's the final one, the one that men also must suffer and perform.

Rebirth: in Western culture this motif is predominantly the preserve of men, thanks in part to the Christian tradition. They are permitted to give birth to another version of themselves. Neo in *The Matrix* swallows the red pill of reality and wakes up naked and hairless in a synthetic, slimy pod, plugged into the Matrix via grotesque umbilical cords. The bonds are cut and our enlightened hero is washed away to freedom via a kind of sanitary birth canal. Elsewhere, in fiction, such a process of purification is not always a success: 'The most racking pangs' heralded the transformation of the good Dr Jekyll into the evil Mr Hyde: 'a grinding in the bones, deadly nausea and a horror of the spirit that cannot be exceeded at the hour of birth or death'.

Dr Jekyll's experiment was intended to free man – in the narrower 'male' sense – from his basest traits and bring forth his noblest form. Instead, the creature emerged 'from the slime of the pit', a failure which Jekyll traces back to his faith in God, which is found wanting. 'Had I approached my discovery in a more noble spirit, had I risked the experiment while under the empire of generous or pious aspirations, all must have been otherwise, and from these agonies of death and birth, I had come forth an angel instead of a fiend.'

But it is not only in fiction that men tried to rid themselves of unwanted weaknesses. In the late 1970s, the German sociologist and writer Klaus Theweleit ploughed through literary sources from the period between the two world wars for his book *Male Fantasies: Women, Floods, Bodies, History* – a classic in men's studies. According to Theweleit, men at the time saw their masculinity defined by what it was not: supposedly foreign, supposedly feminine, supposedly effeminate. Hard and unrelenting, a man's mind was to hold firmly to the values of German culture and 'soul' in order to keep itself from disintegrating under the approaching waves, be it the red flood of ideology from the East or too soft a heart.

In the reality of war, the tides of sump, mud and slime were particularly horrifying. These terms were often used interchangeably by soldiers, writes the British historian Santanu Das, who describes the treacherous terrain of the trenches and long marches as 'slimescapes'. It wasn't just the 'sucking mud' of liquefied earth that trapped and even drowned men, Das writes. The viscously soft ground also harboured metal debris, human waste and rotting flesh. It seemed to have a life of its own, watching like a giant octopus before throwing itself on its victim, surrounding and burying him. These slimescapes were hell, for the body and the mind. Theweleit quotes the author Ernst Jünger, here a young man before an assault: 'Like slime, fear crawls around inside us with a thousand tentacles locking every fibre in its sucking arms.'

Hardened bodies and unshakeable minds seemed only attainable by denying one's own nature from a very young age. Any soft or even erotic element was interpreted by this kind of man as weakness. It was necessary to keep one's feelings under wraps, beneath a kind of armour. But the body was not so easily tamed and contained. There were always those treacherous secretions and fluids, betraying our human nature. 'The floods and stickiness of sucking kisses, the swamps of the vagina, with their slime and mire', they became a physical manifestation of all that was terrifying, Theweleit writes.

Slime still works as a metaphor for everything that can be dangerous, disgusting and simply wrong about sex. One of the most striking examples comes from Vladimir Nabokov's anti-hero Humbert Humbert, who recounts his travels with not-yet-teenage Lolita, whom he serially raped: 'And I catch myself thinking today that our long journey had only defiled with a sinuous trail of slime the lovely, trustful, dreamy, enormous country that by then, in retrospect, was no more to us than a collection of dog-eared maps, ruined tour books, old tyres, and her sobs in the night – every night, every night – the moment I feigned sleep.' But according to this, it's not the girl but the country that suffers.

3

The Stench of Death

'Look at the water. Smell it! That's wot we drinks.'
Charles Dickens, *Bleak House*

Disease is weakness, sex an embarrassment, and death our greatest fear. The experiences at the boundary between health and infirmity, between me and you, between life and the end of it are all slimy by nature. Perhaps there is an unconscious belief that if we push slime to the margins of our consciousness, we might be able to ignore our biological finitude too. We have a collective blind spot. And it is here where slime conceals itself, coating the boundaries inside and sometimes outside the organism and in the environment, but with such restraint that we are able to relegate it to the periphery of our perception.

When we do come face to face with slime – in the flesh, or on the screen – it seems to provoke an automatic disgust response. But why is slime so gross? And what is disgust for in the first place? It is one of our basic emotions and its functions have been argued over by philosophers and legal scholars as well as scientists for a long time. From a purely biological perspective, disgust is an essential protective mechanism that helps shield us from microbes and parasites. 'I define hygiene as the set of behaviours that animals, including humans, use to avoid infection,' wrote the late Valerie Curtis, then a researcher at the London School of Hygiene and Tropical Medicine, and added: 'In humans, the avoidance of infectious threats is motivated by the emotion of disgust.'

But that essential protective mechanism has a weakness: how can disgust warn us of microbes when they are too small to see? How can we stay away from possible contamination if we don't recognize the danger? In order to work reliably, disgust has to take a detour: our early-warning-system of revulsion is triggered by potential symptoms of contamination, phenomena that often indicate the presence of pathogens. What is a common denominator of these wide-ranging microbial threats, covering sickness, sex, death and putrefaction? Slime, of course. Slime is often associated with germs and works well as a trigger for disgust because it speaks to several senses at once. It's quick to reveal sources of infection; we recognize it from a single glance, a single touch.

But that sounds more straightforward than it actually is. Disgust proves to be a universal emotion with a decidedly personal slant: 'It is incredibly powerful, but not based exclusively on logic and reason,' Curtis told me. In her survey looking at triggers of revulsion, British women listed among other things obscene language, women from Burkina Faso mentioned pigs and women from the Netherlands offered fishmongers' hands. These choices often seem arbitrary outside a particular country; there is a touch of local flavour to our aversions. They also show how flexible disgust is, exercised as strongly in response to unpleasant substances as to behaviours, ideas and even words – in short, any kind of transgression of boundaries.

As a defensive response, revulsion can affect whole groups of people, particularly powerless minorities. 'Frequently, these exclusions are underwritten with a rhetoric of disgust,' writes Martha Nussbaum. In Curtis's survey, young Indian women listed contact with people of lower castes as a source of disgust. For Indian people of higher standing, Dalits are considered untouchable in a literal sense. An invisible boundary separates them from the rest of society and often forces them into hazardous work, like cleaning latrines or working in the sewer system. Thus social exclusion leads to contact with excrement, which, in turn, entails rejection by society – and always has. 'The trade of sewage worker was more perilous and nearly as repugnant to the people as the trade of executioner, and held in abhorrence,' Victor Hugo writes in his 1862 novel *Les Misérables*.

What we see is a vicious circle of social stigma and aversions to bodily secretions which operate universally, because fluids such as saliva, sweat and faeces are – locally shaped dislikes notwithstanding – worldwide triggers of disgust. They are followed by corpses and spoilt foods. In another publication Curtis listed the most common triggers for disgust and sorted them into six categories, including poor hygiene, promiscuous sex and visible infection. More diffuse are atypical manifestations like homelessness and other deviations from the social norm, as well as coughs and deformities. This is one more trigger that has been around for a long time: 'There is something wrong with his appearance,' writes Robert Louis Stevenson of the monster Hyde born from Dr Jekyll, 'something displeasing, something downright detestable. I never saw a man I so disliked, and yet I scarce know why. He must be deformed somewhere; he gives a strong feeling of deformity, although I couldn't specify the point.' Curtis's fifth category is wriggling with mice, mosquitoes and other fauna which carry disease and are closely connected to spoilt food, the final class of triggers. How universal are they? In the survey, flies were reported as triggers for disgust in Britain, Burkina Faso and in the Netherlands.

Still, these categories may seem arbitrary at first, but they are linked by a common thread and threat: just as expected, it is the potential for contamination. Slimy sex with numerous partners increases the risk of infection, just like contact with the sneezing sick or the ingestion of mouldy food. These triggers indicate where microbes and parasites may lurk that our repulsed responses then can help us to avoid. Acute disgust will even power up our body's immune defences, which shows their close link to the emotion. Other phenomena are harder to explain: why are women generally more easily and more quickly disgusted than men? It's possible that their propensity for revulsion helps to compensate for temporary weaknesses in the immune system – and not just with regard to their own bodies.

Pregnancy is one of the most critical phases of life, because a woman's bodily defences are dampened so that the foetus doesn't find itself in its mother's immunological crosshairs. This molecular muzzle, however, comes at a time when a toothier form of defence would be helpful, because embryonic development is so sensitive in the first three months. If the mother's disgust is easily triggered, it may well keep pathogens away. A low threshold for disgust in women may also help to protect children in the early years of their lives, when they tend to stay close to their mothers and their own immune systems have not quite reached maturity.

Disgust is a human emotion with deep biological roots. Even other species, such as ants or some crustaceans, give sick members of their species a wide berth, and primates avoid potentially infectious items. If a substance is organic and soft, warm and moist – a little like slime – it will be avoided. This caution pays off because killers can lurk in all bodily secretions. In humans, these can include syphilis spirochetes in sperm, tubercular mycobacteria in nasal mucus, HIV in the blood and cholera vibrios in faeces. The bacterium *Clostridium botulinum* primarily produces its neurotoxin in poorly preserved foods. The potency of this toxin can be seen in Hollywood – so many beautiful faces, so few facial expressions. Injected beneath the skin, a low dose of Botox induces muscle paralysis which can iron out deep wrinkles. Poisoning via spoilt foods, however, poses a risk of respiratory arrest.

The deep biological roots of revulsion mean that we are predisposed to disgust, and in a sense hardwired from birth to develop aversions. We may even be predisposed to be more easily repulsed by slimy and wriggly worms, for example, than by fluffy kittens. But, as all parents of babies and toddlers can testify, humans don't come equipped with specific triggers for disgust; even the wriggliest and slimiest worms are perfectly fit to be touched or maybe even tasted by small kids if no one stops them. In other words: we are born to be disgusted, but have to learn from our parents and environment during our early years what it is we are supposed to be repulsed by. And slime is a particularly potent signal, in industrialized countries at least, a kind of hyper- or uber-trigger, I would argue. That makes biological sense, since slime is in many ways connected to contamination. But there's a catch.

Our avoidance response to slime is lightning-fast, though this speed comes at the cost of nuance. Is it really a biological slime we're cringing at? Is there really a harmful pathogen lurking somewhere within? Faced with possible contaminants, we can't afford to mull it over. Our defence system prefers to play it safe and it will tolerate false alarms. When it comes to avoiding dangerous infections, this appears to be a price the body is willing to pay. Always the mallet never the scalpel, disgust is a blunt instrument and will respond to slime of any kind, regardless of actual risk. As a logical consequence, slime is being forced out of our ultra-hygienic world. But that's easier said than maintained: slime is essential to human and all other organisms. In other words: we can try to ignore slime but it's part of our nature. And there's a downside to it: if we neither recognize nor truly know slime, how are we supposed to appreciate it or use it for our own ends?

Thankfully, even robust aversions are never a hard-and-fast rule and nor should they be. Disgust is an immensely powerful emotion. Ingrained and easily triggered dislikes could be paralysing and ossifying or even damaging if they hinder flexibility. Nature knows how to prevent this, adjusting the threshold of disgust according to our needs. Suspect food, for example, begins to appeal once hunger bites in earnest. Our threshold for disgust also rises in step with sexual arousal, so that saliva, sperm and other fluids no longer seem so unpleasant. After sex, this can swiftly change and keeping your distance from unfamiliar secretions might start to seem like a good idea after all.

Culture plays an important role in the training of our disgust as well. Could we, then, retrain our revulsions, lose or adjust them?

If we are looking for a way to rehabilitate slime and go from unmitigated repulsion to a more nuanced approach, history could give us a clue. Biologically speaking, slime has always been closely associated with disease, but didn't invariably trigger the same kind of disgusted reaction. At least once even the symptoms of a fatal affliction became *en vogue*. Marie Duplessis was regarded as one of the most beautiful women of her time, but her delicate beauty came at a high price. On 3 February 1847, she died of tuberculosis at the tender age of twenty-three. She was immortalized in literature by one of her lovers, Alexandre Dumas, as *La Dame aux camélias*, Marguerite Gaultier: 'In short, either because it was her nature or else an effect of her state of health, her eyes flickered intermittently with flashes of desire which, if spoken, would have been a heaven-sent revelation to any man she loved.'

Dumas was clearly aware of the fatal nature of her beauty but seems not to have been repulsed by it. 'How is it possible that a disease characterized by coughing, emaciation, relentless diarrhoea, fever, and the expectoration of phlegm and blood became not only a sign of beauty, but also a fashionable disease?' asks the historian Carolyn Day in her book *Consumptive Chic*. She has an answer prepared: the symptoms of tuberculosis matched the dominant beauty ideal of the time. Patients lost weight, while a constant, slightly raised temperature made their cheeks rosy and their eyes shine. Ladies cursed with robust health even affected a consumptive air with tightly laced corsets and make-up. The disease only lost its deadly appeal towards the end of the nineteenth century, when the German physician and pioneer microbiologist Robert Koch identified the pathogen which causes tuberculosis and revealed that the disease was infectious. An awareness of pathogens was a turning point in medical and social history.

Before Koch's discovery, though, who or what took the blame for horrible infections? It was not yet slime: first came the stink. In other words, slime wasn't the first phenomenon to be seen as a source of infection and thus triggering revulsion. Before microbes were identified and recognized as potentially dangerous pathogens towards the end of the nineteenth century, contagious diseases were blamed on bad vapours, so-called miasmas. These were thought to be dangerous exhalations in the air; sometimes all odours were even believed to be a threat to human health, one which was hard to escape. Every pocket of still air was thought to pose a potential health risk, as the French historian Alain Corbin explains in his history of odour, *The Foul and the Fragrant*.

'All the miasmas of the cloaca are mingled with the respiration of the city; hence that foul breath,' writes Victor Hugo. 'The air taken from above a dunghill, this has been scientifically determined, is purer than the air taken from above Paris.' How, though, to escape these foul miasmas, which emerged not only from carrion and slurry but the unrelenting earth itself? Paving was meant to protect the citizens of Paris and other cities against this, providing an insulating layer of stone. Hospital beds were designed with long legs to escape harmful vapours close to the ground, while domed ceilings did away with the niches and corners which sheltered sinister miasmas.

But sometimes there was no escape: in 1858, the medically sanctioned superstition, together with extreme weather conditions, brought life in London almost to a standstill. Citizens hid in their homes and the government practically ceased to govern. But the record temperatures that summer, with thermometers regularly hitting over 30°C, were not to blame. The culprit was great mounds of excrement which cloaked the city in a suffocating haze. The city numbered three million people and its outdated and hopelessly overwhelmed sewage system emptied straight into the Thames, as Charles Dickens describes in *Little Dorrit*: 'Through the heart of the town a deadly sewer ebbed and flowed, in the place of a fine fresh river.' As the water level in the Thames dropped, faecal matter and other filth lay exposed on the river's banks, metres high in places.

The Seine in Paris only avoided a similar fate because human waste was no longer flushed into the river, being valued instead as fertilizer. 'There is no guano comparable in fertility to the detritus of a capital,' Victor Hugo writes. 'If our gold is filth, on the other hand, our filth is gold.' In London, by contrast, fears were circling of another outbreak of cholera. Cloths soaked in chlorinated lime, a disinfectant, were thought to ward off the slurry's miasmas and the risk of disease which accompanied them. But sterilized curtains accomplished little against cholera, a devastating epidemic which would ravage London time and again. This was because many households were supplied with water which came directly from the Thames, described by the eyewitness and physicist Michael Faraday as an 'opaque, pale-brown fluid'.

As the historian Rosemary Ashton writes in her book on the Great Stink, it is probable that, at some point, people stopped drinking the disgusting water. Complex emotions are difficult to reconstruct from historical records. But proof of a contemporary response of repulsion on a scale we experience today is rare, if it existed at all. Fear reigned and the miasmas were maligned in very eloquent terms for centuries, including during the Great Stink. But the filth itself?

Powerful people campaigned for sanitary reform in the nineteenth century. Charles Dickens was among them, and he decried in a speech in May 1851 the urgent need for improved sanitation: 'Sanitary reforms must precede all other societal remedies . . . neither education nor religion can do anything useful until the way has been paved for their ministrations by cleanliness and decency.' And he used his fiction to underline the point, as in the opening of *Bleak House* with its portrayal of London as a cesspit of presumably miasmic fog and filth in the streets: 'The raw afternoon is rawest, and the dense fog is densest, and the muddy streets are muddiest . . .'.

Why is it important if fear or disgust makes people act? Fear can be an efficient motivation for change. Disgust tends to be a bit more paralysing, I would argue, at least if it comes to the extreme aversion we feel towards slime and other secretions nowadays. Unlike the population of nineteenth-century London, we don't have much opportunity to confront that dislike. Is it possible that our disgust of slime is a luxury? Is it possible that it can fester to an over-the-top reaction that keeps us from engaging with the material at all, because that repulsion is hardly checked by reality? It seems like it: in a recent publication, it was proven for the first time that disgust has a biological function and can help to prevent infections. It was also shown that the individual environment regulates the emotion to some extent. Populations in highly pathogenic areas without clean water, proper housing and sanitation simply cannot afford to let their most justified aversions run free.

Survival means that we have to live in our world as it comes, and if that entails mounds of faeces or rotting vegetable slime in the streets or 'crust upon crust', as Dickens has it, hardly anyone will be able to bunker at home. This could explain why the inhabitants of London by rights feared for their lives during the Great Stink – and the cholera did return – but looked for the wrong culprit. Fear, and presumably some disgust as well, motivated change in the end but there was nothing paralysing about that repulsion. Not even at the highest level of society: Queen Victoria undertook a pleasure cruise along the polluted Thames, which only came to an abrupt end when the stench proved to be intolerable after all. Such apparent equanimity is hardly imaginable in an age like our own.

And the miasmas? They have been relegated to the shelves of medical history. Their killer image faded as the science of infectious germs became incontrovertibly established in the public sphere. It was then too that consumptive emaciation lost its appeal in Victorian society. The fashion changed and was even put to work in the fight against infections like tuberculosis. Stiffened corsets eventually became taboo. It was agreed that the blood ought to be able to circulate unhindered. Women should be able to take deep breaths of the now miasma-free air. Shoes became the focus of fashion for a time when floor-length skirts were replaced by higher hemlines; this was thought to prevent microscopic pathogens from swirling up off the street and being trailed into people's homes.

Men too couldn't escape the trends unscathed. Traditionally, shaggy sideburns, majestic moustaches and big beards were thought to keep the face warm on the battlefield and avoid the need for shaving with unhygienic razors. But once facial hair was revealed to be a danger to the general public, it had to go. In *Consumptive Chic* Carolyn Day quotes an American doctor complaining of the immeasurable number of bacteria and lethal germs lurking in the 'Amazon rainforest' that is a bearded face. Measles, scarlet fever, diphtheria, tuberculosis, whooping cough and many other maladies were transmitted 'via the whisker route'.

In other words: smells are off the hook as unhealthy miasmas and now serve different functions. An offensive stink still makes us pay attention. A musty smell at home, for example, can indicate an infestation of mould or rotting food, while cadaverines are released during the process of putrefaction. These odours are not dangerous in themselves, but can point to a risk of contamination and warrant a disgust response. Harmless smells, of course, have lost all scent of danger and lovely fragrances even delight us. It is a nuanced approach that lays blame where it belongs and uses symptoms of contamination like disgusting smells as welcome warnings.

Why should we not make our peace with slime as well? So far, in our modern-day response to viscous secretions of all kinds, disgust and a mostly reflexive defence mechanism still reign supreme. We even lack the vocabulary to distinguish slimy friend from oozing foe. 'Mucus' might be the most neutral word for all things slime we have in everyday language, but it comes with a certain medical flavour. 'Hydrogel' is the correct term for most or all biological, but also other, slimes, yet is probably too technical a term ever to become endearing. But if smells can be fragrances, shouldn't we be able to name 'feel-good' slimes as well? Just as not every scent is a deadly miasma, not every slime is an oozing menace, after all.

It doesn't help that the characteristic texture of slime can trigger disgust incredibly fast. Recoiling at lightning speed is helpful when it comes to contaminants, but not a whole class of biomaterials.

In other words, our disgust response is entirely inadequate if we want to do justice to slime as a phenomenon. The complexity and nuance of a thing can rarely be conveyed without precise words, clear definitions and the knowledge of what it is we are dealing with. In the case of slime, this is a group of essential and highly sophisticated substances.

This is where researchers must lend a hand, just as they contributed to the rehabilitation of smell. The spell of miasmas as sources of infectious diseases only broke once the real culprit was identified. We now know that microbes are to be blamed. This is why they were, until recently, in the crosshairs of our suspicion. Now they begin to enjoy a more nuanced reputation, thanks to what we have learnt about their crucial services in the environment and our own bodies. Science leads the way, as the American microbiologist Theodor Rosebury wrote half a century ago, in relation to bacteria: 'If scientists are repelled by the subject, others are apt to be revolted by it. But, with increasing knowledge, revulsion sometimes yields to fascination.'

Smells have escaped their miasmic stranglehold. 'Good microbes' are now all the rage. Slime should follow their example, and science could very well open that door since gooey biomaterials are steadily gaining in appreciation. This book hopes to provide an insight into this still-secret world of slimes, the science behind them, their fascinating richness and their phenomenal significance as biological substances – and to set disgust to one side.

II
Physics

If slime is a phenomenon and an idea, it is first and foremost a physical substance with chemical components and a structure that determines its biological, medical and technical functions. And still defies definition. Slimes in organisms are aqueous hydrogels, water caged in a three-dimensional structure. But not all slimes are hydrogels and not all hydrogels will look or feel slimy. The characteristic consistency of slime corresponds to a specific physical behaviour: slimes are generally viscoelastic, uniting the properties of liquids and solids. This makes them flexible and able to alter their properties as needed. It's no surprise that their uses in nature are extremely varied, including as lubricants and adhesives, tailor-made for the species and organism. Slime is also an essential, selectively permeable barrier inside the human body and elsewhere, allowing in nutrients and other desirable visitors, but stopping pathogens and pollutants in their tracks.

4
Water in Chains

Now the hagfish is a most disgusting animal both in appearance and texture, and some of its habits are nauseating.

John Steinbeck, *The Log from the Sea of Cortez*

When H.P. Lovecraft was creating the fictional coastal town of Innsmouth, cast in bleak shadow, with the menacing Devil's Reef in the distance, he took inspiration from Newburyport in Massachusetts. It is one of many cities on America's coasts to have lost its shine with the end of the golden age of fishing. Even in places like Morro Bay, California, where the industry endures, tourism has become more profitable. Those on the hunt for a dark, looming omen will have to make do with Morro Rock, a round body of solidified lava, the inside of a volcano which has long since eroded away. It stands like a filling in a rotten tooth, throwing only a modest shadow, such that any Lovecraftian horror can only be conjured with considerable effort.

The weather here is too good, the coastline too picturesque, and the wildlife seem to have waltzed straight out of a Disney film. Sea otters play in the waves with their young, herons bask on the beach and seals stretch their plump bellies in the sun. And yet amid the tranquillity of Morro Bay lurks a monster straight from Lovecraft's playbook, as slimy as a creature from Sartre's nightmares. It doesn't get much more extraterrestrial than this. Two hearts? Tentacles coming out of its head? Four rows of pointed teeth? A vertical smile on its face? But would you call it a smile? Or a face? Why, it's the hagfish of course.

The writer and Nobel Prize winner John Steinbeck was not a fan of the monster. He found it 'disgusting' and 'nauseating', while noting that his close friend, the marine biologist Ed Ricketts, 'did not feel this, because the hagfish has certain functions which he found fascinating'. And I find them fascinating too.

The hagfish has certain peculiarities, including its German name, *Schleimaal* (slime eel), which is misleading, since this elongated creature is not an eel – but neither is it a true fish, as its English name suggests. In fact, hagfish, like the parasitic and similarly unpleasant lamprey, are the last remaining members of the primeval cyclostomata, which translates as 'round mouths'. The name is better suited to lampreys, which possess disc-shaped suckermouths with far too many teeth, enabling them to latch on to fish and shred flesh from their flanks. A sensational find confirmed the unique evolutionary journey of cyclostomes. It was the first time a fossilized hagfish emerged, 100 million years old and fantastically well preserved, complete with traces of slime, like a 'sneeze in stone'. It underlined the close relationship between these creatures and lampreys and proved that they are not primitive ancestors of us vertebrates, as some scientists had previously assumed.

Unlike the lamprey, the hagfish appears to be mostly harmless, living in the depths of the ocean where it feeds largely on carcasses. On land it's only seen by those who, like me, patrol the streets of Morro Bay looking for a fish wholesaler. Sandy Winston is one such man, and he takes pity on me on this dull December afternoon just before Christmas and lets me into his yard, where hundreds of hagfish are coiling and contorting in two outsized metal containers which are supplied with a constant flow of water from big hoses. It doesn't stay that way, since the creatures promptly transform the liquid into a stringy, gelatinous mass. I watch it happen and feel it too, because soon I'm rooting around in the transparent slime, trying to grab a hagfish with both hands.

They're harder to catch than I'd expected because they never stay still, slipping through my fingers. And then there's the slime, which is so tough that I can lift it up as a dense fabric. It is so stringy that it creates webbing between my fingers, and so sticky that I won't be able to wash it off, instead having to rub down my hands with a slime-stiffened old cloth, which Sandy's colleague Becky gives me without a word. Blonde and quick to laugh, she stands next to me by the container, rooting around in the slime too. She's responsible for the animals and is now searching for a dead hagfish. 'I can smell it,' she says. Why remove a dead hagfish surrounded by carrion-loving members of the same species? 'They don't eat each other,' says Becky. 'No one eats hagfish. Only Koreans.'

Indeed, their charges are bound for the Asian market, where they are often skinned and placed live on the grill. Korean waters were fished clear of hagfish long ago, so their supplies now come from California and elsewhere, where there is little competition over the catch. Sandy the wholesaler also keeps albino hagfish, which rarely end up in the net, but it's not an attachment to his pets that saves them from the barbecue: it's their horrendous taste. He once tried one under the guidance of one of his Korean buyers: 'I couldn't get it down, even with lots of sauce.'

But their taste hardly factors in life on the sea floor, for predatory fish rarely manage a first bite. The hagfish's loose skin makes it too difficult to catch and there's the extremely slimy defence. Perhaps this is part of what fascinated Ed Ricketts. When threatened, hagfish release extra-long molecules from their skin, stored as space-saving spindles, waiting to be put into action. Then they positively explode, binding vast quantities of water into a dense slime in a split second, forming a suffocating cloud of gel that will even gag a shark. The tens of thousands of fibres in each litre of hagfish slime are long and thin yet durable and elastic, a bit like robust silk or synthetic fibres. And, as entirely natural molecules, they could just show the way towards novel eco-textiles.

But there's more: the US Navy is trialling the use of military lab-grade hagfish slime to stop suspicious ships in their tracks without using force. Current tactics involve launching plastic ropes that slow the ship down by getting tangled into its propeller engine but are hard to untangle afterwards. A weapon based on synthetic hagfish slime might snot a suspicious ship in its tracks instead by expanding underwater into a mass of mucus and dissolving later without residue. It would be the modern recreation of a mysterious slime-like sea described in Greek and Roman antiquity. 'For over 2,000 years, geographical writings have been haunted by mention of a "congealed sea", which prevents ships which reach it from sailing any further or makes the journey much more arduous,' wrote the German historian Richard Hennig in 1926. 'Mention of this phenomenon appears in the Middle Ages too, in tales of the "congealed sea", the "motionless sea", seen from time to time under its Latin name, *Morimarusa*.'

Who knows how many other unique and potentially useful slimes are out there? If the hagfish is the king of animal goo, its kingdom encompasses all of nature – and every single species within it. Biological slimes are not obscure exceptions, they are the rule, and essential for survival. In all my years of researching this fascinating material, I've yet to encounter a slime-free creature and doubt that such a minimalist even exists. Small wonder, as there's hardly a single evolutionary question to which nature has not found an answer in slime. As Mark Denny of Stanford University wrote in a trailblazing publication in 1989, invertebrates in particular rely on the stuff for movement, communication, reproduction, self-defence and even food, while jellyfish, comb jellies and other zooplankton are composed entirely of gelatinous mesoglea.

Spineless or boneless invertebrates are not some dead-end or distant shore of evolution; in fact, they make up nearly 97 per cent of all species. And thanks to their goo-filled lifestyle slime can become more than just a useful material. We need to open our eyes and observe these often small and supposedly 'uncharismatic' creatures to see it. During a severe illness, the journalist Elisabeth

Tova Bailey kept a garden snail, observing and studying it for hours and days on end. 'Slime is the sticky essence of a gastropod's soul, the medium for everything in its life: locomotion, defense, healing, courting, mating, and egg protection,' she writes.

An all-encompassing medium in life: this is what slime means to microbes as well. But what about us, the so-called higher organisms? We vertebrates, as well as plants are not above this. We use slime in myriad ways too, we're simply a little less obvious about it. This is a necessity for terrestrial species, because the highly hydrated slime dries out quickly in the air. Any creature using it on a large scale on land will prefer to hide it away inside its body or as a plant in soil, where water loss can be more readily controlled. The only openly slimy surfaces we humans display are the eyes. They are covered in thin films of mucus which are protected from dehydration by a lipid layer. Slime's mostly secretive nature might be the reason why it escaped our notice for so long as an essential and sophisticated substance.

And what is slime in the first place? A problem, if we're looking for a strict definition to hold on to. This question is central to this book, but surprisingly difficult to answer. It's the default term for unknown but sluggish fluids or creepily soft solids. It's a thing in between and a feeling and a description of materials, but there is not one prototypical slime. Depending on origin and function, it hides behind a multitude of pseudonyms like gel, biofilm, mucilage, glycocalyx, but also in ecological communities like biological soil crusts, or in phenomena like marine snow.

But how to unravel the differences and commonalities here? Most gelatinous substances are lumped together under the label of 'slime', even in scientific publications, without their molecular inner lives receiving much illumination at all. Or at least it was that way a few years ago: today, ever more researchers working on specific slimes are connecting with colleagues to join the dots. An international collaboration under the lead of Dr Adam Braunschweig at City University of New York seeks to investigate different animal mucuses and use their designs to develop new technologies. It's a worthy goal since these slimes are, as they state in a publication, 'remarkably diverse and include lubricants, wet adhesives, protective barriers, and mineralizing agents'. And the brand-new discipline even has a name: 'mucomics'.

But even if specific slimes can be pinned down by their structure and functions, which definition should this book take for all kinds of biological slimes? It's a question I've spent a long while pondering, since I neither wanted cautiously to exclude fascinating materials because they weren't 'slime' enough nor to include substances that factually didn't belong here. Ultimately, I have plumped for a freer interpretation. Many biological slimes have not undergone the research required to know the details of their structure, components or behaviour. But this book is less concerned with the minute details of slimy materials and more with the phenomenon itself in all its diversity, properties and its formidable importance. In short, if it looks like slime, behaves like slime, is regarded as slime or simply catches my attention in a slime-like way, it belongs in this book.

★★★

The definition of slimes may be as slippery as the substances themselves, but they share at least some important characteristics – in regard to their components, structure, behaviour and function. And it might just be easier to unravel this from the outside in, starting with the functions. Varied as biological slimes are, they act mainly as lubricants, glues and selective barriers. There are other functions like hydration or mineralization, but these can often be subsumed into the major categories which themselves are not entirely distinct. They overlap but take into account that slimes frequently appear at interfaces where organisms interact with each other or the environment. Lubricants will allow near-frictionless contact while glues stabilize it. Selective barriers, on the other hand, modulate exchanges at the interface. Many or most organisms produce different slimes whose features and functions can be adapted as needed.

So far there are only a few species whose slimy repertoire we know about and have researched in any detail. Snails, for instance, can crawl along just as easily as they hang from a surface simply by secreting a different slimy glue. And they coat inner surfaces like the digestive tract with slime as a barrier – just as humans and many other organisms do. But what if there aren't any tissues or surfaces inside a body to protect, when one cell is the whole organism? Microbes are the original and possibly the most proficient slimers of all. They bunch together and build themselves a gooey city or biofilm wherever there is some water and surface to attach to. Microbial slimes in the environment are ubiquitous enough to affect habitats from deserts to coasts by glueing sand, sediment and other substrates together, often on interfaces between air, land and water.

Now let's consider behaviour: how can slimes act as lubricants, glues and flexible barriers? This is due to their viscoelasticity, their ability to behave like a fluid and a solid at the same time. In many cases, organisms can adjust this behaviour, fine-tuning the fluidity, the stickiness and density of their slimes, which makes them particularly adaptable to changing needs. Most fluids like water are very consistent in their behaviour. They never become thicker or thinner, for example, while being stirred. By contrast, how hydrogels behave depends on how long and how intensively certain forces are acting upon them. This is mainly what makes them so varied and adaptable as lubricants, adhesives and barriers – despite their being little more than water. The characteristic sluggish fluidity or viscosity depends on the material's inner structure and components.

Biological slimes are highly hydrated hydrogels with water making up to 99 per cent of its mass. 'Slime is little more than stiff water,' according to the German microbiologist Hans-Curt Flemming. Part of that rigidity is thanks to a three-dimensional framework which binds the water – holds it in molecular chains.

In other words: the water wants to flow but is kept on a short, if elastic, lead by the molecular framework, which accounts for the more solid behaviour. This network is formed of polymers, long-chain molecules which are cross-linked. They are unique in their ability to bind together extraordinary quantities of water, at least when it comes to high-functioning slimes produced by biological organisms. Elsewhere, viscous 'slime' can just happen, for example when clay minerals bind an excessive amount of water.

'Slime is more a characteristic than a substance in itself,' says Flemming, and he is right, but biological slimes have more in common. As simple and unformed as they may appear from the outside, their functions and behaviour are based on a very specific structure depending on equally specific components. And it is their seemingly unformed nature that allows a high degree of sophistication and flexibility. If a slime layer is defective, for example, a healthy organism can provide replenishment which easily hooks into the existing gel and closes the gap. A slime matrix can often be fine-tuned to meet its temporary or lasting need.

The gelatinous mesoglea which makes up a jellyfish's body is interwoven with elastic fibres for stabilization, while other biological slimes can be runny on demand. For example, barriers in the female genital tract thin for a few days in the cycle to let sperm through, at least the ones with excellent timing. A gel barrier might be equipped with antimicrobial agents to keep pathogens at bay, while snails leave molecular messages for potential mating partners in their trails – which might just as well attract hungry predators.

It seems that all life forms are slimy, in one way or another. And yet we know very little about it. This is partly due to the fragile nature of biological hydrogels: stiff water tends to dry out rather easily and the highly variable molecular network proves to be elusive as well. Science simply lacked the means to study these substances in detail until now. Recent years have seen new tools, methods, projects and collaborations in an effort to understand and use hydrogels better, with some astounding results. But even if we are only now able to analyse hydrogels in detail, it is not the first time other slimers have aroused human curiosity and been exploited for human purposes.

The yellow slime of the banded dye-murex snail turns to deep violet in bright sunshine. Thousands upon thousands of these molluscs were killed in Ancient Rome, and later also in the royal households of Europe, for extravagant gowns in imperial or royal purple. The common piddock *Pholas dactylus* had the similar misfortune of being extraordinarily prized. These creatures grind holes in stone using their elongated shells, concealing themselves inside for life. Despite their nifty burrows they were found by the Romans, and Pliny writes of night-time *Pholas* feasts during which guests' mouths, hands and clothing would glow in the dark because they had been sprayed with the molluscs' bioluminescent slime. It looks like we might just ooze our way back into a future where we use and appreciate nature's gels in even more surprising – and hopefully more subtle – ways.

5

Stuck on Slip-ups

[I] ran my index finger over a slug and tasted a little of the slime. The result was splendid; first I began gesticulating a little like an ostrich, secondly, I had to drink a glass of Cognac and, when that didn't help either, a glass of bitter, and then another, and then, thirdly, I lost my appetite for three days and, fourthly, the affections of a very pretty girl to whom I, in my incredible folly, communicated the incident . . .

Hermann Löns, *A Disgusting Creature*

The wedding took place on 20 September 1881: Victoria of Baden, a German princess, wedded Crown Prince Gustaf V of Sweden and Norway, becoming queen at his side years later. However, despite the couple's three sons, the marriage only existed on paper: Gustaf was gay and Victoria spent a significant amount of time with her lover in Italy. When she was back home in Sweden, she had a summer residence, Solliden Palace, built on the island of Öland. No detail was too small for Victoria, neither in the palace, which became a homage to her Mediterranean home, nor in the surrounding park, where she had the large red slug *Arion rufus* imported from Baden and released into the garden.

Unfortunately, the species felt right at home and spread widely. Even in neighbouring Norway it is now considered an invasive pest which could have been avoided if Victoria had been content with a closely related species in Sweden that also fulfilled a purpose, alas not a decorative one. In pre-industrial Sweden and well into the twentieth century, the black slug *Arion ater* was put to use as a lubricant on carts. As the Swedish ethnobiologist Ingvar Svanberg writes, children collected the molluscs, which were kept in their own containers in the carriages. When the dry axle began to squeak, a slug would be placed on the wheel hub and crushed as the carriage continued on its way. This practice appears to have been widespread; the German writer Hermann Löns also mentions that freight drivers used the slugs as cart grease.

Maybe it was the long-plotted revenge of the slugs that led to an accident on the motorway outside Paderborn, Germany, in the summer of 2016. A Trabant skidded on a wide trail of slime which had been left behind by a nocturnal mass-migration of snails. For all the havoc they can wreak, these phlegmatic creatures never experience similar catastrophes in their own world – and not just because they live life in the slow lane. They don't slip off course on wet tarmac, or gritty sand, or slippery mud, remaining in constant close contact with the ground. They are able to conquer any seemingly impassable terrain, both on land and in the water, or to stick to it, thanks to their slimy repertoire of lubricants and adhesives.

'How can an animal with only one foot walk on glue?' Mark Denny wonders in a publication of 1980. We now know that the proverbial snail's pace is compelled to be so by pedal mucus, which is so powerfully adhesive – even when serving as a lubricant – that the creature is hard-pressed to free itself from the slime, and the slime in turn is difficult to remove from the ground. However, a snail's slime will rub off if its foot pulls the viscoelastic hydrogel in one direction while the ground exerts force in the other direction via friction. This complex mechanical stress is called shearing and sees two parallel forces exerted in opposing directions. In daily life, this can occur when a car skids. Just like a snail, the car pushes ahead while the roadway decelerates backwards. This conflict comes at a cost to the car's tyres, which are worn down between the two competing forces.

Snails don't produce rubber burn, but they do leave a trail of slime. This kind of crawling is viscoelasticity on an intermediate level, and snails have perfected the method by generating their own surf, allowing them to ride the waves in slow motion. A series of contractions sends numerous little waves across their soles. Elisabeth Tova Bailey observed this in her pet snail: 'My snail secreted a special kind of slime for locomotion, called pedal mucus, over which it traveled. While its ability to glide over a patch of moss appeared effortless, when it went up the glass side of the terrarium, I could see bands of minute ripples moving across the underside of its foot.' Every elevation presses directly on the slime below, causing its molecular framework to break or shift if the pressure is strong enough. The snail foot then glides on the more fluid slime. Once the wave has passed, the pressure abates and the molecular framework recovers. The slime hardens again and the snail foot pushes away until the next surge.

It is a highly specialized form of locomotion which requires the snail to keep lifting its foot as if pulling it out of sticky chewing gum. This restricts the pace, takes time, strength and additional energy, because slime – that precious resource – is left by the wayside. Crawling on slime is a real extravagance, one in which snails invest a third of their total energy. All other kinds of movement in the animal kingdom, like hopping, running, swimming and slithering, for example, are a comparative bargain when it comes to energy use. Even flying can be cheaper. In other words, economy matters, particularly on porous ground, where garden snails (*Cornu aspersum*) are obliged to produce even more pedal mucus. An athletic gait can help avoid the need to splurge. The snails arch their feet over the ground and push themselves forward like caterpillars. Some scientists refer to this as hopping, though the creatures do not move any faster. It's not merely a matter of pace, but budget. These speckled skinflints keep material costs down by leaving individual dots of slime instead of one long trail.

Other energy-saving snails either can't or don't want to get into these kinds of gymnastics and opt for recycling instead. They use the trails of their fellow snails, only replenishing the slime as required, while possibly boosting their love life. The males of some species of winkle recognize potential partners from the slime trails, which they read like detailed lonely-hearts ads. Information about the other snail's sex, species, direction of movement and even attractiveness is all contained in its slime. Well-fed ladies with broad trails free of parasites are particularly popular, it seems. Advertisement-by-slime is helpful for slugs and snails; after all, they rarely meet by chance, and speed-dating is out of the question.

And it works – as long as the slime stays true. Male rough periwinkles (*Littorina saxatilis*) are far greater in number than females, and a clutch of eggs laid by a single female can be fertilized by more than twenty males. If the suitors' pursuit becomes too persistent, embattled females might leave behind a sex-neutral slime trail. It's an act of subterfuge with unexpected consequences and leads male suitors up the garden path. In no other species have scientists been able to observe more attempts at copulation between confused males who have accidentally wooed each other.

★★★

The message is always in the mucus: the rosy wolfsnail (*Euglandina rosea*) eats other gastropods, tracking them by their slime trails. To do this, the wolfsnail has converted an offshoot of its lip into a sensory organ, a kind of handlebar-like slime scanner. The predatory mollusc follows almost any trail, preying on land snails of all kinds, and it's highly efficient about it: 'If you put four Partula snails and one Rosy wolfsnail together, nothing will remain of the Partula snails after 24 hours,' writes Florian Werner in *Snails: A Portrait*.

Fellow members of its own species, with whom a Rosy wolfsnail may come to mate, have it slightly easier – providing they don't find themselves part of a scientist's experiment, smeared with the mucus of potential prey snails. In these situations they are mercilessly attacked by their stalker, while prey snails smeared with *Euglandina* slime have found themselves unwittingly playing the role of potential mating partners.

Marine molluscs must also adapt to unpleasant visits. Some bladder snails, for instance, constantly coat their bodies with mucus to lubricate their way on and into the sediment, trailing a delicate sleeve of slime behind them. Unfortunately, this slime can attract the attention of predators, including the elegant sea slug *Navanax inermis*, which creeps on slime itself and will add bright-yellow deterrents of an unknown nature if under attack, to keep even bigger hunters at a distance.

Leaving your mark in mucus can be dangerous, but also extremely helpful. Some limpets in the tidal zone need it for guidance. At first glance, their homes appear to be little more than unremarkable indentations in the rock. But the edge of any home scar fits perfectly around the shell of the resident limpet, protecting it from drying out when the tide ebbs. But how do these animals find their way back to their own scars in time? A limpet's slime trail serves as an external, spatial memory, leading it through the labyrinthine tidal zone like Ariadne's thread, glistening but not too obvious. In the 1940s, the Californian researcher Willis G. Hewatt pondered this still-mysterious behaviour. Without knowing about the mucus trail, he scored a deep groove across one unlucky snail's homeward path, as author Rachel Carson (1907–1964) describes in her wonderful *The Edge of the Sea*: 'The limpet halted on the edge of the groove and spent some time confronting this dilemma, but on the next tide it moved around the edge of the groove and returned home.'

Female owl limpets (*Lottia gigantean)* are not only at risk from the low tide and the scientists; they have reason to fear young members of their species too. These snails are all male at first, only later turning into females, with a switch in behaviour as well. Older owl limpets live as recluses and, like many limpets, defend their own patch in the tidal zones. And the stakes are high. Over many tidal cycles a layer of pedal mucus will gather around the home scar, and the female inhabitants cultivate a garden of nutritious microalgae on the slime which may even act as fertilizer. The snails keep competing algal growth down and graze in moderation on the vegetation. This is when unruly teenage limpets can become a nuisance. The young males lack the same green-fingered ambitions, preferring to poach food from other sites. Things will get nasty if they come across the rightful owners because the latter use every means possible to combat greedy neighbours and thieves. They bulldoze intruders out of the way and are ready to come to blows. A limpet's home scar is its castle, and mucus is a vital part of the fortress.

Limpets not only find their way back via slime or live off it; they also use it to seal themselves off while in their home scars. Unlike the pedal mucus used for gliding, this is a special sticky slime that many tidal but also marine and even terrestrial invertebrates secrete for endurance. They can glue themselves firmly in place, whether underwater, in the surf, on even upside down from a tree. It is only now that we are starting to understand how complex these adhesives really are. The common limpet (*Patella vulgata*) stumped scientists for over a century because it can cling so tightly to rocks in the tidal zone that it becomes almost impossible to remove. Surely, strong muscles must be responsible? Actually, it is sticky slime and a pioneer at that, because it was the first biological glue to have its 171 proteins catalogued in detail.

Much to the researchers' surprise, they encountered old friends here: molecules that had been found in similar form in other biological glues from starfish, sea urchins, flatworms and other invertebrates. It seems that nature relies on basic principles and a set of ingredients to concoct a myriad high-power adhesives that are still adapted to individual needs. Starfish, for example, move on hundreds of feet that need to stick with every step before the glue immediately dissolves again. Mussels, on the other hand, choose a spot and settle down for good, never to come unstuck again, and barnacles even cement themselves to a substrate. If many or all biological glues have common components, it might be easier to imitate them in the laboratory. And the hunt is on for novel adhesives, since conventional synthetic glues rarely work on damp or bloody surfaces or even underwater, and if they do they're often toxic.

One medical innovation at least is already modelled on snail slime. *Arion subfuscus* is another slug that princess Victoria might have added to her collection. It uses its slime to anchor itself so securely to the ground that not even hungry birds can pull it away. American scientists took structural inspiration from this strong glue for a new surgical adhesive, far superior to standard products. It works in damp and wet conditions, and has even been used to mend a hole in a pig's heart, tolerating many thousands of contractions. In the meantime, it has withstood another test: spina bifida is a developmental disorder that leaves the spinal cord and brain exposed. While the neural tube can be closed after birth, prenatal surgery would protect the developing foetus much better. But where to get a non-toxic adhesive that tolerates foetal movement and growth and sticks even while submerged in amniotic fluid? In an animal model, at least, the slug-inspired glue worked even better than expected.

Even recyclable adhesives might come from nature's slime armoury, maybe by way of the Australian red triangle slug (*Triboniophorus graeffei*). At first sight it seems to have escaped an attack with a razor blade that left a blood-red triangle on its back. In fact its main predator is the tree frog, which hunts with a gooey tongue, as we will later see – and the slug is far from helpless. Instead of sticking themselves to one spot, however, these animals prefer to turn their glue on their attacker. It is a phenomenon noticed by scientists who encountered a rather unfortunate tree frog clinging to a branch within reach of one such slug, but it wasn't eating it and nor could it hop away. The slug's mucus had left the frog helplessly stuck to the branch for days, and even human hands could only free it with some effort to save it from drying out or being eaten itself.

This animal adhesive, however, is another candidate for a novel glue, one that can be reactivated with water after losing its grip.

These are but a few examples of slimy lubricants and glues in nature, but biological adhesives are fast becoming an important research focus. Beyond adhesion, slime has many other functions, and it might not be easy or desirable to prise them apart. The mucous coatings of the digestive tract, for instance, lubricate our food's path, stick to excretions and act as a selective barrier. And this is just one of the different yet interlocked hydrogel systems that shield all functional units in our body, from organs and tissues down to the single cell and its nucleus. Our health and ultimately our lives depend on this delicate shielding of stiff water finding the right balance according to location and often-changing needs. Nutrient versus pollutant, dangerous pathogen versus resident microbe, sperm or cancer cell versus immune cell: who's allowed inside the body, tissue or cell? And who has to stay outside?

6
Shifting Boundaries

But the story of the beautiful sleeping briar-rose, for so the princess was named, went about the country, so that from time to time kings' sons came and tried to get through the thorny hedge into the castle. But they found it impossible, for the thorns held fast together, as if they had hands, and the youths were caught in them, could not get loose again, and died a miserable death.
Jacob and Wilhelm Grimm, 'Sleeping Beauty'

Operation Overlord began at dawn on 6 June 1944 with the landing of the Allies in Normandy. They would face bitter clashes with German troops as they fought their way inland. Here, in *bocage* country, they encountered an unexpected adversary. 'The Normandy hedgerows, enclosing small fields and bordering every road and track, were at least three times the height of their English equivalents, heavily banked and far too dense for even a tank to smash through,' writes the historian Antony Beevor. Help came in the form of Rhino tanks, hurriedly kitted out with steel tines, capable of cutting through banks and hedges in a matter of minutes. Nonetheless, there was a surprising amount of resistance for a boundary built not from stone and cement but woven from living branches. 'The hedge is threshold, boundary land,' writes the literary researcher Ian Rodwell in his blog *Liminal Narratives.* 'It delineates, marks and divides. Poised between field and field or meadow and lane, it signifies the boundary it simultaneously enacts.' In nature, hedges are porous and pose no barrier to air and water, nor to insects, birds or other small creatures. Yet they are impenetrable to wild and grazing animals, and to people too – even fairy-tale princes, when necessary.

In *The Natural History of the Hedgerow*, John Wright speculates that hedges are an age-old tradition and may have helped in early Britannia to keep grazing cows out of fields of grain, while both had to be protected from wild animals like wolves. What they signify is more liminal threshold than absolute boundary. Here lies their commonality with slime barriers, which are just as selective in what they filter out, keeping friends in their place and foes away. To take this analogy further: our body is cultivated and rich land which must be kept free of pathogens, while domesticated microbes are only allowed at a distance.

Hydrogels can be dense but are never completely closed off; they contain too much water, held together by little more than a three-dimensional framework. Both components play an important part in the functioning of the barrier and in its selective capabilities, as we will see in this chapter. Much depends on the building blocks of the network. But the first questions are: what is that highly co-ordinated thicket of molecules capable of, and what does their adjustable selectivity look like? Katharina Ribbeck has been researching these mechanisms at her laboratory at the Massachusetts Institute of Technology (MIT) for years in an attempt to understand mucous barriers at a more comprehensive level.

According to her work, any particle's ability to pass through a hydrogel depends at first on its size and the width of the mesh that forms the slime framework. Tiny particles diffuse into hydrogels relatively smoothly, like insects passing through a hedge. Expansive particles above a certain size, however, simply get stuck, whether the thicket be botanical or molecular. The slime acts as a sieve, allowing for simple yet effective selection of large contaminants and pathogens on a purely mechanical level.

But size is not the only factor; the molecular framework can select in other ways. Some unlucky particles pass the sieve test but will nevertheless find themselves entangled in the gel. The electrostatic interaction between their surface and the gel framework causes them to become caught, as though snagged on thorns in a thicket. The converse is possible too: if a large particle is lucky, these same interactions will prove key to safe passage through the gel.

However, the distinction between friend and foe isn't always easy to make. If our hydrogel barriers are not taken into account during the development of a new drug, for example, the medication could be intercepted and fail. This means that any therapeutic agent encountering a slime barrier must carry with the correct permit, so to speak. The Rhino tanks' method – brute force – will do more harm than good here, as breaches in the barrier will also allow harmful pathogens to slip through in the wake of the drug. It is for this reason that biomedical research is currently focused on molecular Open-Sesame-style mechanisms, making hydrogel barriers penetrable for certain agents but enabling them to snap shut afterwards.

Certain pathogens already employ some of these mechanisms, having adapted to our gel-based burglary system over the course of their evolution. They may be carrying with them enzymatic wire-cutters which cut a hole through the molecular thicket of the hydrogel. Others proceed more elegantly and copy the code which guarantees passage through some gels. And that is not so very different from what a human spermatozoon must achieve to penetrate the mucus plug in the cervix. Success depends also on the right consistency of the mucus, which the female organism fine-tunes during each monthly cycle. For sperm it's all about timing: many kings' sons had already come, and had tried to get through the thorny hedge, but they had remained sticking fast in it, and had died a pitiful death, as it says in 'Sleeping Beauty'.

★★★

So it's only right to be scared: 'I don't wanna go, who knows what's waiting for us out there,' cries the bespectacled spermatozoon Woody Allen in the comedy *Everything You Always Wanted to Know About Sex* (* But Were Afraid to Ask)*. Like a prince attempting to reach the sleeping princess in her overgrown castle, sperm run the gauntlet on their journey towards the egg, advancing like unwelcome foreign bodies first through the acidic environment of the vagina and then through the mucus plug which safeguards access to the womb. This barrier is almost always too thick and dense for pathogens, and sperm too. This changes during ovulation, when a mature ovum is present and can be fertilized; pH is a crucial factor here – the more acidic the environment, the thicker the mucus.

With ovulation, the pH rises. The hydrogel in the cervix softens by absorbing excess water, it thins and becomes more penetrable. Then, and only then, are at least healthy sperm permitted to enter on their quest to find Sleeping Beauty: '*But by this time the hundred years had just passed, and the day had come when Briar-rose was to awake again. When the king's son came near to the thorn-hedge, it was nothing but large and beautiful flowers, which parted from each other of their own accord, and let him pass unhurt, then they closed again behind him like a hedge.*' It is a complex mode of selection, but it's not just the female body that knows how to rig the game. The man's seminal fluid can also raise the pH of the vagina, creating a more moderate climate for Woody and his cohort, who would otherwise find themselves stuck in the deadly environment outside the gel barricade, surviving only briefly. But what happens when the whole system fails?

Premature birth remains the leading cause of death in newborn babies and can effect severe complications for those who survive. Infections in the womb play a significant role in this. For a long time, however, it was unclear as to how the pathogens found their way into the uterus since access is blocked by the mucus plug, which is supposed to remain impermeable throughout the entire pregnancy. A study done in Ribbeck's lab showed that the barrier wasn't thick enough in many women with a high risk of premature birth, instead remaining as porous as it would be on fertile days. Any pathogens present would have an easy job of it unless science finds a way to strengthen the weak barrier and protect the unborn child. Maybe an as yet uninvented synthetic mucus that would have uses with other bodily barriers as well would help.

Unlike the genital tract, the stomach was thought of as sterile for a long time. What pathogen could possibly survive and settle in the highly acidic environment of an organ whose walls are protected by a double layer of compact slime? It seemed unfeasible, but the bacterium *Helicobacter pylori* manages to do just that by enzymatically neutralizing the stomach acid in its immediate vicinity. It creates a mellow buffer zone whose altered pH opens up a pathway of thinning mucus in the barrier for the pathogen to swim through right up to the stomach wall. It's an old trick to use our own biology against us, it seems, since traces of *Helicobacter pylori* were even found in the mummified Stone Age ice man, Ötzi.

Secreted slimes like the mucus in our stomachs are ubiquitous in animals with a diverse range of properties and functions in the body. They all contain a large proportion of so-called mucins. They are the major components of the hydrogels' inner three-dimensional networks that bind the water. Mucins are very large, complex and – in evolutionary terms – very old molecules. A recent publication on secreted mucuses states that most higher animals express at least five individual mucin genes but can contain more than twenty; in humans, we know of twenty-two so far.

It's the structure that makes a molecule a mucin. Whether found in humans, fish, amphibians, molluscs, birds or one of many other organisms, all mucins look alike, a bit like a bottlebrush. A single protein stretches out to form a kind of backbone, with some areas of the molecule carrying a large number of specific building blocks. These domains are characteristic of mucins and mucin-like molecules and they act as designated docking sites. Hundreds of sugar chains attach here in high density, hence the look of bristles on a handle. But these so-called glycans are a bit more complex than thin bristles: they're long and branched chains made up of different sugar molecules. They stick out from the protein backbone and can in some mucins make up nearly 80 per cent of the molecular mass.

How do cells cope with a bunch of giant bottlebrushes that dislike each other in a tight space? The sweet thicket of glycans makes mucins so bulky and unwieldy that the handling of these gigantic molecules is not entirely clear yet. One model focuses on Mucin 2 (MUC2), which builds the framework for mucus in our digestive tract. It is produced and secreted by so-called goblet cells, a name that refers to their uneven shape, with a swollen upper half crammed full of nicely packaged mucins. But how are these molecules parcelled up? According to the model they team up in threes. And only in the goblet cells' special environment will they suspend their molecular antipathy. Once new mucus is needed, the molecules are released from the cell and the shackles of enforced friendship. Now they can't get away from each other quickly enough, unfurling and extending their dense bristles at an explosive rate – maybe a little like automatic umbrellas. In milliseconds, they can take up water and swell 3,000 times to form a hydrogel which blends seamlessly into the existing mucus.

This hypothesis has yet to be proved conclusively and the glycans' role as part of mucins is still a long way from being fully appreciated. They're essential for binding water and forming gels, but their significance goes beyond the realm of slime. These sugars are a central cog in the organism: alongside proteins, lipids and nucleic acids, they are one of the four major building blocks of life, but the least understood of them. For a long time, there were no methods to study these complex structures even though they look deceptively simple at first. Glycans in mammals are based on building blocks of no more than nine simple sugars. They are being combined in multiple ways into long chains – just as about twenty amino acids make up all proteins, or as a set of distinctive beads would be enough to create ever more necklaces with unique patterns. That means that in the human body alone, those few simple sugars create several tens of thousands of different complex glycans.

As if that weren't variety enough, these chains attach themselves in various and changing combinations to mucins and many other proteins but also lipids and even specific nucleic acids, as a recent publication has shown. About that last association very little is known so far, but what is certain is that protein and lipid partners need their glycans; without them they misfold, lose stability or can't function. Modern glycobiology might still be in its infancy compared to other disciplines. But thanks to new methods and approaches it is well equipped for the future. It is hoped that it will finally help to explain what exactly these multifaceted sugars do and what role they play in diseases. It's a matter of deciphering whole glycomes – that is all the sugars in a cell, an organ or organism – both during good health and sickness.

Just one example: cystic fibrosis is a condition where great quantities of exceptionally sticky mucus accumulate in multiple organs. In the lungs it provides an ideal breeding ground for pathogenic bacteria, irreparably damaging the airways. In organs like the gut it impairs the uptake of nutrients, while it might make it hard or impossible for patients to conceive, possibly due to a defective barrier in the genital tract. But the lung is hit particularly hard in these cases and its mucus is insufficiently hydrated.

A mucin involved in this barrier shows a starkly altered and unique glycan pattern. That means, it displays a combination of sugar chains that has only been found in patients with cystic fibrosis so far and might change the molecule's function, including its ability to bind water. This in turn could prevent the mucus barrier from being fully hydrated. If we understand glycans better, we might be able to help patients with this and other diseases. A deeper understanding could also pave the way for a unifying theory of biology and possibly herald a new era of modern medicine.

As we have seen, MUC2 is secreted by goblet cells to build the framework for extensive gel layers in the digestive tract, as do the majority of our mucins elsewhere in the body, in the airways or genital tract, for instance. But the secreted mucuses that coat our inner surfaces are just one type of hydrogel system in our organism. There are three more, and one of them also contains mucins, if not the gel-forming ones that are secreted by goblet cells. These mucins are not fully released from their cell but remain anchored to its membrane as part of a team of proteins and lipids that stay attached but reach out to the cell's environment.

Most or even all of them carry glycans, with just a few to tens of thousands of sugary building blocks per chain. They form a dense layer of sugars that covers the cell's surface. Interwoven with them are additional free glycans and together they are the so-called glycocalyx – literally 'sweet husk' – of the cell. It is a universal feature, as a recent paper states: 'Every cell in the human body – endothelial cells, immune cells, muscle cells, blood cells, neurons, and all others – exhibits a glycocalyx,' writes the researcher Leonhard Möckl at the Max-Planck-Institute for the Science of Light in Erlangen, Germany.

He adds: 'The latest research has shown that the glycocalyx is an organelle of vital significance, actively involved in and functionally relevant for various cellular processes, that can be directly targeted in therapeutic contexts.' In his view the sweet husk is a 'fundamental cellular agent'. But is it also a slime? Only to a certain extent. A robust glycocalyx might rise high from a cell's surface and resemble thick woods in profile but it is not a regular network. There is some structure and connection, but no three-dimensional framework as in mucus. But still, as Möckl mentions, the first identified function of the glycocalyx was protection, and this is thanks to its function as a physical barrier for any object to enter the cell: 'The glycocalyx is a dense, gel-like meshwork that surrounds the cell.'

Millions of glycans can make up the unique glycocalyx of each cell. It works a bit like a molecular barcode to help our immune system; for example, to differentiate between friends, like regular body cells, and microbial foe. As with treacherous slime trails, being easily identifiable has a downside too: viral pathogens, for instance, enter host cells that they target via their telltale glycocalyces. In fact, the cellular glycocalyx has received a lot of attention lately. The coronavirus that caused a pandemic in 2020 docks on to human cells that carry the heavily glycan-adorned receptor ACE2. This molecule is found on the surface of many different cells from the heart to the gut, lung, kidney, brain, testis and blood vessels – which makes the Severe Acute Respiratory Syndrome coronavirus 2 (SARS-CoV-2) so much more dangerous.

How can that be remedied? Blocking ACE2, which is indispensable for different cellular functions, seems impossible. But maybe the virus could be lured from its destructive path before it docks on? In some labs around the world synthetic mimics of the cellular receptor are now in the works, a soluble ACE2-decoy that could potentially be taken up to fool the virus into assuming it had found a target cell while never coming close to any tissue, only to be removed by the body's defences. It's an elegant and presumably old trick: other molecular receptors that are routinely targeted by dangerous viruses when they hit our inner surfaces are thought to be secreted as decoys into the extracellular mucus as well.

Mucins might be among them, but they also fight back in different ways. The membrane-bound variants on inner surfaces such as the digestive tract are thought to be able to release their outer part once a pathogen is attached. And their size seems to matter as well, when the towering mucins shadow underlying receptors and protect them from any pathogen's grasp. Another stage of tricks and betrayal features Sleeping Beauty and her undaunted prince: sperm are surrounded by a thick coating which protects them on their perilous journey. But identification counts here too, and their sweet gel might prove single sperm cells as suitable candidates. Egg cells wear an elaborate coat of glycans which acts as a last hurdle for any lovelorn spermatozoon, proving him as a healthy candidate from the right species. A real prince, not a frog stuck in disguise.

III

Organism

There is probably no slime-free life form in existence and maybe there never was. Human beings are not special or different in this regard; we're simply the best-researched multicellular organism when it comes to slime. Our bodies are equipped with four different hydrogel or gel-like systems, so that invading pathogens encounter a new slimy barrier at every level. These are highly specialized gels, adapted to particular needs with a multitude of functions, and they co-operate with the immune system. The extensive slime barriers which coat our internal interfaces, such as the digestive tract, also interact with billions of resident microbes which live inside us and contribute to our well-being. This symbiotic community is so tightly interlinked that it's hardly possible to view animals and plants as individuals. Each multicellular organism forms what is known as a holobiont, the assemblage of a host and the microbes living in and on it, and slime makes that coexistence just so much more peaceful.

7

The Human Fortress

Oh calm down, calm down. All right what was it?
There was . . . the lights . . . and the phone . . .
and . . . THE WALLS WILL OOZE GREEN
SLIME? Oh, wait, they always do that.'

Squidward, 'Graveyard Shift',
SpongeBob SquarePants

'In all the days of my life, I have seen nothing which touches my heart so much as these,' Albrecht Dürer wrote in his diary of the 'wonderful things for human use' which he encountered in Brussels in August 1520. They were the treasures of Moctezuma, the last ruler of the Aztecs, which had been plundered by the Spanish conquistadors under Hernán Cortés and given to Charles V, the Holy Roman Emperor, and exhibited in Europe for the first time. Golden suns, silver moons and other intricate works of art, weapons and sumptuous textiles were displayed to a distinguished public. Dürer's interest in the Aztecs endured, such that, years later, he intensely studied a print of their magnificent capital of Tenochtitlán, which was circulating even in his home town of Nuremberg, and incorporated elements of its layout into his own work.

It was an age of envisioning the future. Thomas More, Lord Chancellor to King Henry VIII, had recently published his *Utopia*, whose fictional isle gave a name to the notion of an ideal society. In 1527, Dürer made a bid of his own; he wanted to prove his extensive learning and abilities in theory and practice as the *uomo universale* that he was. With *Various Lessons on the Fortification of Cities, Castles and Localities*, Dürer sketched out his own stony utopia in fine detail. Military security was indispensable, but building fortresses was no longer enough: defence should instead fuse with the civil community, bound by strict aesthetic principles. It was an ideal city, planned down to the slightest detail, mathematically perfect, like a Nine Men's Morris board – and just as boring. Like so many utopian designs, it lacks the charm of an organically grown structure. In More's words, 'He that knows one of their towns, knows them all.' But what does the ideal city have in common with the human organism and its attendant slimes?

The parallels lie in the challenges which must be overcome as much in stone as in flesh, blood and mucus. How do you protect a site which is under constant threat from foreign armies or microbial hordes without interrupting the essential processes of life and survival, such as procuring food? Dürer's fortress city boasts a strict hierarchy; the castle is the control centre, safe and secure at the heart of the city. Potential usurpers will have to get past a number of streets and several levels of defence or find a back door. Housing and places of work, shops and stables form functional units. It reflects the social order but also allows for efficient division of labour: the blacksmith's forge is next to the foundry, the clergy live close to the church and the town hall and the noblemen's homes are close to the castle. The higher principle, however, remains defence; the unending crush of invaders leaves its mark on city and man alike.

Within the human body, our civilian side keeps the organism going; the cells are its smallest unit, and – depending on function – combine to create tissue and organs, which then form groups as closely connected systems, like the digestive tract. Here, as in Dürer's city, things that belong together usually group together. Yet this stringent order goes beyond the mere arrangement of organs. The bowel must combine defence against pathogens with the intake of nutrients, two contradictory functions which it approaches in very different ways. The tissue of the alimentary canal, which at first glance appears uniform in its structure, is divided into segments. Initially its focus is on taking in nutrients, while the immune system takes a back seat. Pathogens are only countered aggressively in the lower sections of the intestine, where robust defence is unlikely to hinder digestion.

In areas where the receipt and supply of nutrients to the organism have priority we afford pathogens a little more leniency, but never too much. Vital processes are only allowed to continue undisturbed if access to the body is still under strict control. And we have four complex and interlocked hydrogel systems as barriers to help enforce it. Where exactly is the first line of defence? It might just be a dose of healthy suspicion. Like a fortress city, we are constantly on the alert for foreign aggressors, but also internal saboteurs. Dürer presented potential invaders with a multi-level system of ramparts and ditches to keep invaders at bay before they reached the city itself. This is not unlike our own first line of defence against pathogens. Scientists refer to a 'behavioural immune system', in which the emotion of disgust plays a key role. In other words, shying away from people who might be infected or from contaminated material might keep us in health. And what if everyone, including yourself, could be carrying a dangerous virus? Then we just have to strengthen the ramparts by pulling up our drawbridges, keeping safe distances, avoiding handshakes and wearing a mask.

But it's not always possible to avoid all contact with pathogens, and the body's physical defences must come into play. Our skin is our bulwark, a constant reminder of the body's integrity. In Ancient Greece the shell of the human body was revered despite the somewhat ignoble nature of skin, which was seen simply as the outer layer of the gelatinous and sticky body underneath. According to the *Book of Skin* by Steven Connor, in Ancient Egypt anyone was 'an object of general hatred who applies violence to the body of a man of the same tribe or wounds him'. As a consequence, embalmers could only work on bodies that had been incised already. That made the 'slitter' an indispensable man, and yet no less hated, it seems, since after making the incision on a dead body he would be obliged to flee the scene at once, lest he be cursed and pelted with stones by onlookers.

Modern-day surgeons are less reviled but broken skin is still dangerous. Our outer layer is a dense barrier and allows for little cross-border traffic; its purpose, first and foremost, is to regulate water loss. A biological city wall of sorts, it seals itself with a dense layer of dead, keratinized skin cells stuck together. Pathogens are hard-pressed to settle in this dry, nutrient-poor, acidic environment. Not only that, but the skin is already colonized by friendly, specially adapted microbes, most of which will not suffer rivals. Entry to the organism via the skin is frequently only possible through open wounds: a breach in its defences. No wall can be completely secure, though; every fortress city must interact with its environment, be it simply to build up its reserves. And this is what the skin allows with our gateways, including the mouth and nose.

Structurally speaking, these openings represent the junction between the impermeable skin and the porous mucous membranes, which make up all internal surfaces that come into contact with the outside world and coat them with hydrogel. They cannot seal themselves off like the skin does and it is this which makes these interfaces, and the whole organism in turn, vulnerable. Invaders land here in enormous numbers. According to a study by Stanford University, more than 1,000 million microbes can find their way into the body every day simply in the meals we eat.

Not all germs pose a risk, but the microorganisms in spoilt food and their attendant toxins certainly do. Every bite is rigorously tested by the body's gatekeepers – our senses. Is it slimy, does it smell or taste in some way suspect? If so, it is rejected. At least, most of the time. A range of traditional delicacies, such as the Japanese dish Nattō, are so fermented – undergoing a process of microbial degradation – that they develop a stringy slime. Those who haven't learnt to love the scent and texture of such specialities in childhood may struggle to overcome their own disgust response if they encounter them later in life. Lovers of Nattō might be equally repelled by the slimy French dishes of escargot or frog's legs, or by a bowl of gloopy British porridge. Rancid flavours or putrid smells require a similar familiarity in order to make them palatable, as with the stinky Asian fruit durian, a delicacy which makes itself known from miles away, or the whiff of Parmesan cheese, which is sometimes (and only by the uninitiated, of course) likened to old socks.

We register scent molecules and flavours when they land on receptors in our nasal cavity and on our tongue, at which point they must be dissolved in slimy saliva or in the mucus inside the nose. It is one of the first critical tests at the entrance to the human fortress. Alain Corbin writes of 'the sense of smell as an instrument of vigilance . . . the nose, as the vanguard of the sense of taste, warns us against poisonous substances'. This function might be important enough to make it team up with the tongue, where scent receptors have also been found. In this way, information about taste and scent merge before they reach the brain.

We mustn't overstretch Dürer's city as an analogy, but both there and inside an organism defence is staggered across multiple levels. For stony fortresses, this would force foreign invaders to negotiate moats, storm enormous walls and strike down or outwit the site's defenders. In our bodies, four hydrogel or gel-like systems form their own, tightly interwoven slime barriers which perform both civil and military functions, securing interfaces which are also lifelines. Hydrogels must ward off intruders, secure supplies for the community in the form of respiratory gases and nutrients and maintain internal order. Those who do not contribute have no place here, nor do they in Dürer's city: 'The Lord shall not allow useless people to live in this fortress, but talented, high, wise, manly, experienced, skilful men, good tradesmen who are capable with respect to fortifications, rifle-makers and good marksmen.'

In our bodies, pathogens and parasites are the most obvious undesirables, but tumour cells also refuse to play by the rules.

They evade control, growing and spreading without licence to do so, using the body's restricted resources. Mercenaries, on the other hand, are most welcome, but they must be kept in check. They form the microbiota, a troop of all kinds of microorganisms trillions-strong. All these different kinds of microbes including bacteria, fungi and so-called archaea, mainly populate the mucus layers and are indispensable for our well-being.

The focus of research, for the moment, is on bacteria, though these are trumped in number by viruses. Their community, the virome, or phageome, does not pose any danger to us because it is composed of bacteriophages, viral predators that only attack bacteria. These viruses are stuck to the framework of our mucus, where they accumulate and wait for their victims. They check the numbers of our bacterial colonizers and ensure a healthy, diverse microbiota, but also strike down pathogens which might be passing through.

The first tier of our physical defences are extensive layers of gel, a kind of moat which floods our gateways and thoroughfares with slime. Microbes which find their way in on the air we breathe, the food we eat or during sex become stuck, are filtered out, washed away or put into a deep sleep. But it doesn't always work. If germs manage to swim through the gel in the digestive tract, they come up against the mucous membrane itself underneath. The cells here are densely packed to prevent any microbes from slyly slipping through. Like a Roman phalanx, they carry a robust shield that is the sum of their individual glycocalyces. As with all other open and vulnerable tissues, the cells' dense gel coat is particularly thick. Here, lots of giant mucin molecules, among others, do their very best to protect the body's frontier.

This is all part of communicating in glycocode: All of the body's cells sport a glycocalyx, a tailored suit of sugary armour that helps the immune system tell the difference between our own cells and pathogens or other enemies. However, in the long evolutionary arms race with our cells many microbes have learnt how to develop appropriate counter-measures. Harmful Group B streptococci, for instance, simply clothe themselves in human glycans so as to avoid exposure. A similar tactic is employed by cancer cells that assume fake identities by manipulating their own 'tumour glycocalyx', which will help them grow and spread while looking harmless to the immune system.

This is a case of deadly leniency, but errors can be made the other way too if our body's aggressive reaction towards a foreign glycocalyx turns out to be a sort of friendly fire. This can happen in its reaction to transplanted tissue or a blood transfusion. To let these cells fly under our defensive radar, scientists are now trying to create 'blank' blood cells without any telltale outer markings for universal transfusions. For this they're using a mucin-cutting enzyme that comes from a gut bacterium that's wielding this powerful hedge-cutter to shear a path through the dense molecular thicket. Pathogens which don't know how to do this might just find themselves entangled in their target cells' glycocalyx.

Until recently, the sugary layer was mostly seen as a glorified buffer zone. 'We now know that the glycocalyx is functionally involved at the core of cellular events of high relevance for both health and disease – from membrane organization all the way to cancer progression,' writes Möckl. We are just beginning to understand this sophisticated barrier that is also a communication hub and interface between cell and environment. We do know, though, that a defective cell coat can be dangerous. As open-faced tissue, the walls of our blood vessels carry a thick glycan coat. It saves the cells from being shredded in the bloodstream while also transmitting signals to them so that they can react to any changes. If the blood flow is interrupted during surgery or as a consequence of pathological crises like systemic inflammation, the glycocalyx can shed. This might make the walls of the blood vessels permeable and lead to other serious defects, or even to organ failure.

★★★

To return once again to the image of the fortress city, an individual's access was determined by the kind of work they performed; stable boys had no business in the castle, for example, while law enforcers of all kinds could move about the community quite freely. Our body's immunological protectors patrol with the same freedom and carry identification via glycocalyx. Almost all other cells, though, are restricted to a neighbourhood, their respective tissues or organs. And what belongs together, glues together: cells of a tissue interact via their glycocalyces but are also anchoring themselves to another hydrogel. This glue is the extracellular matrix, or ECM – our third hydrogel system in the body. It is better known as connective tissue and literally holds us together by creating three-dimensional tissue from individual cells. Without it, there would be no organs, no muscle fibres, blood vessels, bones, ligaments, tendons, cartilage or skin. Or blood, which contains different cells and other components – suspended in liquid ECM.

'Many diseases – in fact most diseases – are probably ECM diseases,' the late Zena Werb argued, having spent decades researching this hydrogel and its role in breast cancer. The extracellular matrix clearly plays a part in infections and cancer, controlling incoming and outgoing activity at the body's borders; and there are not only foreign invaders to fear. If tumour cells want to travel and grow elsewhere, they not only have to tailor their glycocalyx into a passport, they also have to detach themselves from their traditional connective tissue and latch on to a foreign matrix environment. That is easier done by cancer cells than understood by science: the ECM is highly specific for each tissue and usually only hospitable to cells that truly belong there. So how do cancer cells fit in?

Biomedicine hopes to unravel the mysteries of this campaign of conquest because most cancer deaths are caused by metastatic tumours that can spring up far from the primary tumour – but not just anywhere. In 1889, Dr Stephen Paget developed his seed and soil theory of metastasizing cancer cells being as reliant on a suitable environment as seeds on fertile soil. This concept hoped to explain why metastases do not appear at random, instead – depending on the site of the primary tumour – afflicting only specific organs and tissues. Still, for a long time science focused primarily on the complex molecular and metabolic changes in tumour cells as if they were free-floating agents and removed from their environment. Now, the host tissue gets a place in the spotlight as well, and it seems that migrating cancer cells do not simply conquer new habitats. They also need a suitable microenvironment of supporting cells embedded in an altered extracellular matrix.

Microbes, on the other hand, must first conquer the slimy moats and glycocalyces of the mucous membranes – only to find themselves equally stalled in the extracellular matrix. This so-called basement membrane is a tough layer of gel directly beneath the cellular surface of tissues and organs. Like the impenetrable thicket around Sleeping Beauty's castle, this tightly woven network pervades the walls of the human fortress. Here immune cells are on patrol, and unlike the fairy tale's sleeping soldiers they are wide awake. This might make it sound much more uniform and passive as a barrier than it is. The extracellular matrix takes many shapes and forms in one organism, specific for each tissue or organ, a three-dimensional environment perfectly adapted for the cells embedded in their respective connective tissue.

It's perfectly adapted in the sense that this hydrogel keeps cells on the right track by talking them into the correct behaviour. Molecular signals coming from the ECM tell immature cells to differentiate into a specific cell type, to grow, to move and to arrange a diverse team of them into complex structures like bone or blood vessels. If we ever learn how to talk ECM, we might be able to usher in a new era of biomedicine by growing even organs on demand, by printing them on 3D printers or – better yet – by boosting the organism's own regenerative powers to replace them. It might be impossible to replicate the human ECM in all its diversity of shapes and forms, as it is composed of more than 300 components in varying combinations.

It might be enough, though, to go for a pared-down version, to develop minimalist yet functioning ECM-scaffolds or to use natural stand-ins that do the job as well. First attempts and successes, such as 3D bio-printed corneas for implantation in the eye, have already been recorded. Also promising is an approach that uses new kinds of ECM-like hydrogels, some of them extracted from edible seaweed and furnished with highly specific signalling substances. These materials can be injected into the body, where they seem to speak enough ECM to recruit all necessary cells to the site and initiate the formation of complex three-dimensional structures like blood vessels.

Moats of slime, gel-like suits of armour and walls overgrown with matrix: only few pathogens will surmount these three hydrogel barriers. Those that do, however, still might not yet be in reach of their goal. Viruses, for example, must capture the machinery of the cell nucleus to seize the host's control centre and use it for their own reproduction. They must then overcome a fourth and final hydrogel. Unlike Dürer's castle in the heart of his ideal city, the human organism is governed in a decentralized way even if all of its parts co-ordinate. Each cell nucleus contains its own genetic material, can give its own commands and is shielded from the rest of the cells by its own protective coating. Here, import and export occurs via thousands of pores, each dispatching up to 1,000 molecular deliveries per second. Who can check that amount of cargo traffic? Well, there is a high security lock of sorts, a hydrogel barrier inside each pore that helps to control that sensitive border.

It's a tough barrier to breach but there are a few specialists waiting in the wings, armed with sophisticated tools and preparing to invade. One of these microbes is a tank-buster; too big to fit through the nuclear pores, it blasts a hole in the nuclear membrane instead. Others show remarkable restraint by waiting for a phase in the life of the cell when the nuclear membrane dissolves of its own accord, clearing a path towards its genetic material. Then there are tricksters like the herpes viruses, little Trojan pathogens which use fake papers coolly to sneak their way in. They activate the molecular shuttle which melts the hydrogel in the pore for wanted cargo, but here helps invaders to slip through like harmless travellers.

It's an evolutionary arms race between our defence mechanisms and the many attackers who seek a way into the human organism. We have to continually up the ante. But what about lowering our defences for a change? One very special parasite relies on the help of its host to throttle its own bodily defence. This is the human embryo nestled within the female body, which not only nurtures it but also protects it by dampening its own immune system. As befits a parasite and its host, it's a slimy kind of connection via the umbilical cord which can bend but not kink because its blood vessels are embedded in connective tissue. Known as Wharton's jelly, this extra cellular matrix collapses after birth to pinch off the blood vessels. During pregnancy, though, it stabilizes the lifeline from mother to embryo. In the future, that kind of support might work in reverse as well, from child to adult: Wharton's jelly contains a great number of cells that can develop into different types, which might be of therapeutic use in diseases from neurodegenerations to diabetes.

8

An Arms Race with Microbes

Later, though, he couldn't resist taking another look in the microscope.
You bastard, he thought, almost affectionately, watching the minuscule protoplasm fluttering on the slime. You dirty little bastard.

Richard Matheson, *I am Legend*

The royal rooms are decorated in dark art nouveau style. Black walls gleam with slime like scar tissue, a metallic whirlwind unfurls: it is the Xenomorph. The queen is busy adding another slippery egg to her clutch. An encounter with the dark queen in Ridley Scott's film *Alien* – with her spindly arms, skeletal tail and elegant neck shield – is unlikely to be forgotten in a hurry. The Swiss illustrator H.R. Giger dreamt up the creature, but possibly took inspiration from another nefarious Supermom. *Phronima* is her name, and she seems imposing from up close with pincers and deep-set eyes like Giger's queen. In reality, though, this marine crustacean measures only a few harmless centimetres in size – harmless for us, but *Phronima* floats in the dark, open sea with her brood in a transparent barrel that was once a creature itself.

Phronima is a parasite and preys on salps, eating their innards while leaving the hollow shell as a nursery. Salps are seemingly simple animals – so-called tunicates – but so closely related to vertebrates that they can help us better understand some aspects of our own evolution. Many tunicates look like gelatinous barrels where seawater streams in through one opening and out the other, propelling the creature along. Any plankton coming along is filtered out by a mucous mesh and digested.

The life cycles of most tunicates are immensely complex, with solitary and colonial phases in which thousands of salps, for instance, form chains metres long. Other tunicates embed themselves together in a gelatinous matrix, forming pyrosomes or fire salps up to twenty metres in length, which can gleam with bright-blue light. The tubes' openings are large enough for a human being to swim through, but divers are advised against it. The matrix is extremely tough and rarely yields. Fish have been known to become caught in pyrosomes, and one fire salp is even said to have been spotted trapping a drowned penguin.

Elegant appendicularians, or larvaceans, on the other hand, look a bit like tadpoles enclosed by their own intricate slime filters, built into giant constructions like fine-pleated ruffs. The creatures' undulating tails pump a steady stream of seawater through the laminated chambers into a larger, slime-filled space to concentrate nutritious particles in ever-greater density. Tunicates belong to a multitude of groups of marine animals which use a sticky net of slime to trap prey in the water and even select them according to size and shape.

The bacterium *Pelagibacter ubique* seems to be destined to make up a big part of any tunicate's diet; as the name suggests, it is particularly abundant, potentially even the most numerous microbe on the planet. In the water column it accounts for about a quarter to a half of all microbes. Easy prey? Not so fast. Thanks to a special surface layer these bacteria seem to be able to slip through even highly selective mucous nets – showing but one example of the evolutionary arms race between animals and microbes.

Nonetheless, tunicates clog their mucus houses quickly with other plankton, often gobbling the whole mess up before producing another sophisticated filter. For some appendicularians, this can be as many as forty houses a day. A sizeable number of these mucous constructions and their planktonic load sink to the sea floor, providing essential ecological services. The blessing from above is an important source of nutrients for other species in the deep but can also stay there for a long time. This is an essential part of the marine carbon cycle, about which there will be more later: gelatinous zooplankton like tunicates incorporate enormous amounts of carbon from their planktonic prey. That carbon comes from the atmosphere, and without long-term storage on the sea floor it might end up there again and, in the form of a greenhouse gas, accelerate global warming.

The story of these gelatinous feeding machines trapping particles to eat has another fascinating angle: they are in some aspects similar to vertebrates and can tell us more about our own evolutionary history. The model of choice for one particular study was the sessile and ubiquitous vase tunicate *Ciona intestinalis*, with a focus on its gut. In evolutionary terms, our digestive systems have always marched to a different beat, because only mammals coat their intestines exclusively with mucus. Invertebrates use a material that is best known from insect and crustacean exoskeletons as a barrier, chitin. Not to be outdone, though, *Ciona intestinalis* combines both.

As a first step, the tunicate filters planktonic particles out of seawater and coats them in a slime which twists into a rope that drags them into the gut – including dangerous stowaways. These are microbes that might not all be as harmful as the bacterial 'bastards' in Richard Matheson's *I Am Legend*, but they still pose a serious risk. If they get a chance. But invading microbes in *Ciona* will find themselves up against a wall of chitin that shields the tissue. Tests have shown that without this physical barrier, the tunicates die. But chitin doesn't do the trick alone; its fibres are embedded in a slimy matrix containing mucin, much like the gels in our intestines.

The similarities end here; our mucus is much better equipped to fight pathogens compared to the simple tunicate matrix. It is a whole new level of sophistication: our mucus contains molecular decoys to bait and trap microbes. It keeps helpful microbes that do not concede territory to pathogens and train our immune system to maturation. Lastly, unlike the *Ciona*, our digestive tract is divided into successive sections where the uptake of nutrients and defence are co-ordinated as needed.

The entry point for food to the human organism is the mouth, a cosily warm and wet environment for microbes. They are met by slimy saliva, a complex biofluid that helps to taste the food, starts to digest it and lubricates its further travels. But it also contains different gel-forming mucins that employ a range of fascinating behaviours to defuse pathogens. MUC5B, and especially its sugary glycans, sends the cavity-causing *Streptococcus mutans* bacterium to some sort of hibernation, an interaction that keeps it from settling on our teeth. Microbes like these usually have to mob together to become dangerous. Maybe keeping them apart is all it takes to put them to sleep. Other mucins, though, choose a less subtle way to remove the risk and trap the microbes to dispose of them in the body's own shredder. This is the stomach, with its vat of hydrochloric acid.

The stomach makes quick work of most microbes and would soon dissolve its own tissue too if it weren't for the double layer of protective slime. The inner layer is an extremely tough gel, set beneath a second layer of less dense slime. Still, only a skilled survivalist could surmount such an obstacle. Yet the bacterium *Helicobacter pylori* neutralizes the acid in its immediate vicinity and melts a path through the slime, gaining access to the tissue and taking hold there, where it is known to cause stomach ulcers and even cancer.

The next stop is the small intestine, which has to deal with an acidic mixture of fragmented food and aggressive digestive secretions from the pancreas; despite this, it makes do with a light hydrogel. If the gel barrier fails, autodigestion can cause fatal organ failure within hours. The slime itself is not only a physical barrier but is also equipped to fight pathogens directly, containing potent antimicrobials and quickly transporting away any harmful cargo. In this moat, pathogens have to be quick to use any chance they get, withstanding the gel's adhesive power and swimming against the deluge of slime. Moving in vicsoelastic mucus is always a challenge to overcome – and to study. Human sperm, for instance, have long been thought to undulate snakelike through the barriers in the female body, but actually they are rolling as they swim, much 'like playful otters corkscrewing through water', as one scientist put it in relation to the findings in a recent publication.

Tiny microbes have it even harder in mucus but will not be deterred. *Vibrio cholerae* is a bacterium that rows against the slimy current with the help of a rotating appendage while at the same time mowing down the thicket of mucins. Others, like the diarrhoeal pathogen *Campylobacter jejuni*, are themselves engine-powered corkscrews – they even swim faster in a viscous fluid like mucus than in water. And the microbe must have other tricks up its sleeve; if it is too slow and carried along with the slimy current, it will try its luck in the large intestine even though the mucus there looks totally different.

On a superficial level, the coating of the large intestine resembles the two-tier system of the stomach in disguise: both organs have a layer of viscous slime adjacent to the tissue, which is covered in another, less compact gel. So how is one pathogen able to attack both sections of the intestine, which are inhospitable in their own ways? The answer probably lies in the gels' molecular frameworks, all of which are based on the same major building block, the MUC2 mucin.

But apart from the structural commonality, the large intestine is very much its own organ, though it long went unrecognized for its service to the human body – a bit like the appendix, a seemingly vestigial and ultimately useless organ. In Victorian England the colon was considered a banal, dispensable tube which had only served a purpose when human beings were still threatened by big predators. A theory claimed that, if danger arose, excretions were held back rather than being expelled, so that predators would not be attracted by the scent. The otherwise brilliant British surgeon

Arbuthnot Lane even went so far as to say that the fermenting mixture of digested food in the bowel could poison the human organism from the inside out, and – without further ado – removed his patients' large intestines.

In one respect, at least, Arbuthnot Lane wasn't completely wrong: the large intestine is a kind of reactor in which microbes take care of remnants of digestible food and ferment it for us. They are part of the microbiota, trillions of bacteria, fungi and other microorganisms which can also be found elsewhere in and on the human organism but are especially numerous and active in the large intestine. They help to defend against pathogens, and one way in which they do this is by leaving no room for their competitors to colonize.

This is not unlike the aggressive wren and other nesting birds in Nigel Hinton's *The Heart of the Valley* that will defend their prime locations:

> The morning after their nest had been destroyed, the two dunnocks set off in search of a new nesting-site . . . On the way back down they briefly flitted along the inner branches of the privet hedge bordering the garden, only to be harried by a tiny but fearless wren. Indeed, all the best sites already belonged to other birds and they grew accustomed to being chased away by the indignant owners.

Be it a botanical thicket or a molecular one, once all the good, or at least available, spaces have gone, new arrivals find it hard to settle.

And in both scenarios, already established parties support each other. Hedge-dwelling birds of different species might warn one another when foxes or cats are near, while some bacterial species provide their microbial neighbours with valuable metabolic products, such as vitamins and energy-rich fatty acids. As the host organism, we also benefit from these and other resources and contributions. The microbiota has a wide-reaching influence on our development and well-being. In return, the large intestine offers food and shelter, taming this miniature menagerie and keeping it under control with the help of its mucus.

It is a balancing act for the large intestine to keep pathogens away while allowing wanted microbes strictly controlled access. There's one area, though, where harmful and helpful microbes are treated exactly the same: the impenetrably compact slime layer on the intestinal wall is additionally infused with antimicrobial substances to keep microbes away, be they friendly or not. Our mutualistic relationship notwithstanding, the vast number of microbial cells still pose a permanent threat to us and become dangerous when they get too close. This is where the two-tier mucus system comes in handy: the hostile barrier right next to the tissue should be off-limits, but the lighter gel on top of it provides a safe distance – and therefore perfect accommodation for our microbial guests.

When it comes to biodiversity, our gut is an ecosystem like any other. A richness of species conveys a certain robustness and helps to withstand disturbances. What about our microbial composition? A typically Western lifestyle with a fibre-poor diet and high standards of hygiene, as well as very effective antimicrobials, seems to be detrimental to our microbiota and its diversity. It is still not clear what makes a healthy and well-rounded microbiota: is it about specific species that prove to be indispensable? Or is it only about certain functions that have to be fulfilled, no matter who tackles them? It could also be a combination of the two.

Even if all microbes were equal in regard to our health, some would be more equal than others – and *Akkermansia muciniphila* probably reigns supreme in the gut. The bacterium is a landscape gardener to our hydrogels and bodyguard to our bowels, managing the colonization of the intestines with a seemingly villainous touch. This microbe eats the mucin sugar chains inside the intestinal slime, but instead of destroying the barrier, stimulates production of fresh hydrogel. This stabilizes the barrier, just as thick hedges benefit from regular trimming. 'The natural inclination of a hedge is to become a wood,' writes John Wright in *The Natural History of the*

Hedgerow, 'so if no management is carried out at all the hedge will just keep growing; the shrub layer will be lost and the entire hedge will become leggy and thus gappy too.'

By stimulating its host organism to churn out more mucus, *Akkermansia muciniphila* secures its own supplies. If the bacterium is well stocked it can spread and keep pathogens away. The rest of the ecosystem benefits from the metabolic products this microbial slime-muncher provides. As do we: the gut bacterium is associated with protection from excessive weight gain and better wound-healing, and it might even help with cancer therapies. If this species suffers, it's bad news for our metabolic health. Resulting conditions include obesity, chronic inflammatory bowel disease and others.

As a consequence, *Akkermansia muciniphila* is being discussed as a possible probiotic, a live microbe for consumption to boost an all too often impoverished gut flora. But how can the whole gut flora be restored when a severe bout of diarrhoea, for example, wipes it out? Another misunderstood organ will come to the rescue: the seemingly pointless appendix probably keeps in store a sample of our microbial diversity to repopulate the intestine in all kinds of emergencies.

The interactions between lifestyle, microbiota and mucus are far from understood, but we know that there are no one-way streets. While the microbes might to some extent dictate our appetites, we try to keep them in check by accommodating them in mucus whose functions they in turn modulate. What does that microbial mucous house look like? And how do the two structurally different slimes come about? Since all mucins are being secreted by goblet cells in the gut wall, the source of the loose overlaid layer is hard to explain. That is, unless both variants turn out to be one and the same mucus after all – just in different phases of its life. The hydrogel starts out as the dense antimicrobial barrier coating the tissue. Its thicket of mucins is so dense that parasites like the amoeba *Entamoeba histolytica* need enzymatic wire-cutters to get through.

It's an old trick that we master ourselves, but we prefer secateurs over a hedge-trimmer. That way – instead of breaching the barrier with brute force – we work the surface of the barrier with such precision that the network of mucins remains intact, albeit in a different form. The network is more relaxed, in a sense more coarsely knit, creating a still-solid but much looser gel. And a completely new habitat emerges: the trillions-strong microbiota makes its home in the upper layer, but, just like pathogens, it is only able to penetrate the bottom layer in exceptional circumstances. The quarantine works as long as the dense slime layer underneath remains intact, chemically equipped with its antimicrobial ingredients. When does the barrier fail? A breakdown can be mediated by pathogens but also by the microbiota when regular sources of energy fail.

'Fight them or feed them' is the title of a recent publication on the dire consequences of a dysfunctional relationship between our intestinal mucus and the microbiota that can threaten the integrity of the barrier's framework by consuming the glycans of mucins. It's all sugar, after all. That happens when they need to find alternative sources of energy because our Western-style, fibre-poor diet doesn't provide enough nourishment. Not every single breach of the barrier will have serious consequences; some low-level inflammation is likely to occur all the time at the intestinal walls. But a sustained attack, when microbes might even penetrate the tissue, will trigger a massive response from the immune system. If the balance of the mucosal ecosystem, with a healthy and diverse microbiota as well as a proficient mucus barrier, is not maintained other immune malfunctions like food allergies can follow as well.

Gels in these expansive frontier zones have to keep pathogens away and the microbiota – an extension of ourselves – at a distance, but they do receive help to do this. They form part of what is known as mucosal immunity, a branch of the body's defences that is associated exclusively with the mucous membranes. Goblet cells, for example, produce mucins, but also work shifts as sentries, passing particles from the mixture of digested food to immune cells that are waiting just below the surface of the gut wall. They patrol the tough basement membrane of the extracellular matrix and will, if they perceive a potential risk, spring into action by calling for back-up or ramping up slime production.

So, is the body's defence system a ruthless killer or a modest manager of microbes? It's both; mucosal immunity must tolerate both its defenders and friendly strangers like the microbiota in an atmosphere of peaceful coexistence. This only succeeds when the course is set early in life. Newborn babies' undeveloped immune systems only mature as microbes begin to colonize their bodies, when the body's defences and microbiota are able to communicate and begin a lifelong conversation.

These precisely calibrated interactions between multicellular organisms and their microbes are still little understood. Yet they speak to a long and joint evolutionary history. All multicellular organisms play host to a dense microbiota and are so distinctly shaped by this that it almost seems pointless to continue to speak of individual organisms. We form an inextricably linked community with our microbes, of which there is at least one for every cell of the human body – we are holobionts.

9

The Illusory Individual

There were horrendous, dramatic, violent quantities of green slime – oodles of it. It covered Howl completely. It draped his head and shoulders in sticky dollops, heaping on his knees and hands, trickling in glops down his legs, and dripping off the stool in sticky strands.

Diana Wynne Jones, *Howl's Moving Castle*

The Hawaiian bobtail squid *Euprymna scolopes* has a problem. It stands out clearly against the moonlit sky as it leaves the sea floor to hunt in the waters around the coast of Hawaii at night. That might make it easy pickings for predators swimming below and scanning for dark silhouettes of prey animals above – if it weren't for the squid's cloak of invisibility. Shortly after hatching, these creatures collect *Aliivibrio fischeri*, a species of bioluminescent bacteria, by trapping them in mucus. The microbes find their home in a pouch in the squid's mantle, a special light organ. At night, the tiny tenants shine brightly enough to hide their host's silhouette from below and get food in return. But it's a short-lived relationship for most of them. Each morning almost all the microbes get kicked out of the light organ; the few left behind will replenish their numbers during the day.

As holobionts, all multicellular organisms form a tightly interwoven entity with their species-rich microbiota. They co-operate so closely that it is almost impossible to tell whether we allow the microbes to settle in and on our body or if they indulge themselves by keeping a multicellular host. Over the course of our long evolution as a community the interactions between us and the microbiota have become too complex to be picked apart easily. But that is what scientists have to do in order to understand the whole process, and that is what makes the squid and its microbes so interesting: it is a co-operation between two partners only.

On the other end of the spectrum are simple organisms that give insights of a complete microbiota. The freshwater polyp Hydra is a good example and is already an established animal model in laboratories, thanks to its extraordinary much-envied capacity for regeneration, which allows whole polyps to regrow from tiny fragments. Hydra might eventually throw some light on our own evolution as holobionts, or at least the development of the hothouse of microbial activity in our organism. The polyp shares some similarities with the human gut, but as an inside-out version.

Like the wall of the large intestine, the outside of the polyp is coated in a tiered mucus, roughly split into two specific layers. A closer look reveals a well-known arrangement, with a microbe-free mucus layer next to the tissue and another coating on top of that – where the microbiota lives. It is one slime layer divided into barrier and habitat, in both the large intestine and the polyp. Nothing new, then? A quick reminder: gut cells sport a thick glycocalyx, a sweet cloak, as a robust shield to keep all microbes at a distance. The two-tier mucus barrier is separate from that and lies on top while being divided in two itself. The glycocalyx and the secreted mucus are two different gel systems, after all. In Hydra, however, it's not a secreted mucus that handles incoming pathogens as first line of defence and accomodates the settled microbiota. It's the glycocalyx that play-acts the mucus barrier of our gut yet stays firmly attached to the cellular surface.

It seems that this kind of divided and strictly controlled interactions with microbial friends and foes must have developed over a long time. The polyp's two-tier glycocalyx might represent an early stage of mucosal immunity for this fought-over interface, while ours came much later. It seems that evolution found the habitat-cum-barrier structure enticing as an organizing principle, with less emphasis on the actual hydrogel doing the job. And that principle demands that we develop a more nuanced approach to microbes. We still need to keep pathogens away, while hugging our microbial friends at a distance. They are part of ourselves and have been for a long time, as the evolution of extended mucus layers suggests.

Slime-coated surfaces have existed in higher organisms for hundreds of millions of years, and some scientists consider it a milestone or even a prerequisite for the evolution of true multicellularity. A true multicellular organism is not just a loose collection of cells, it has sophisticated tissue and organs. Early examples and pioneers of extensive slimes include comb jellies (*Ctenophora*) and cnidarians, such as corals, jellyfish and sea anemones – all of which boast true tissue. By contrast, sponges (*Porifera*), which are an even older life form, might secrete some mucus but lack extended high-functioning mucus barriers, as well as tissue, organs and even a real boundary between their organism and the environment. Their cells are embedded in a gelatinous matrix without much higher organization. Taken apart, sponge cells will scramble to reconnect and are able to build – or is it rebuild? – a whole organism.

Sponges are not frozen in time, but their simple organisms might offer a glimpse into the early evolution of animals. There is some debate over how sponges developed, one suggestion being that they may descend from aquatic unicellular organisms. These hypothetical ancestors are thought to have resembled the ocean-dwelling choanoflagellates of today. How would we know? The most important cell type in a sponge is the *choanocyte*, which looks like the twin of a choanoflagellate. Both trap organic particles in slime and benefit from teamwork.

Choanoflagellates exhibit a striking behaviour when they sense chemical signals emitted by their preferred prey bacteria in the water: they divide and produce daughter cells but don't release them. They will stick together via their extracellular matrix and individual gel coatings in the so-called rosette formation – and hunt much more efficiently as a team. This could testify to the existence of a common ancestor, from which two branches extended: one led to the unicellular choanoflagellates with their tendency to form a colony, the other led to the multicellular sponges, with no possibility of unicellularity since their choanocytes can no longer detach from the organism at will. This theory is disputed, and less differentiated ancestors are discussed, but one point probably stands: microbes might have played a crucial role in multicellular evolution.

Specific bacteria initiate a rosette formation of choanoflagellates while other microbes make up almost half of the sponge's mass. All multicellular organisms grew up in a microbial world, and we needed a facilitator for our relationship with them – which is slime. Unlike wizards such as Howl in *Howl's Moving Castle* by Diana Wynne Jones, we can't produce it at will if a particularly bad moods strikes us or the need arises. Without our mucuses and their multitude of functions, especially in our interactions with microbes, we would be lost – or reduced to sponges.

Even these simple animals sport mucus-like systems and a microbiota, but they lack a true hydrogel-coated interface with their environment. Microbes can penetrate their bodies more or less unhindered, carried in on seawater. Without a high-functioning and tightly controlled mucus barrier sponges might find it much harder to defend themselves against pathogenic assaults. This might explain why they produce such powerful antimicrobial agents. It might also explain why they never developed any true tissue or organs. In contrast, the evolutionary younger comb jellies and cnidarians employ extensive mucus layers to control their interface with the ocean. To some extent they are sealed off, able to keep the numbers of invading microbes in check while providing access to a hand-picked microbiota.

Do we have to thank slime barriers for our very existence? At the very least for the existence of our gut and its unexpected offshoot: the lungs. A protrusion from the digestive cavity is thought to have transformed into a respiratory interface for breathing. This transformation facilitated life on dry land and, over the course of evolution, it created our lungs too. We even have eyewitnesses: the lungfish is the oldest living ancestor of four-legged vertebrates. It breathes air and possesses cells containing hair-like outgrowths. These are known as cilia and move as one, sweeping slime out of the airways. In the lungfish *Protopterus aethiopicus* these can be found both inside the gut and in the anterior section of the lung, which underlines the evolutionary kinship between the two organs.

We carry the legacy of this with us today. Though cilia are not present in the human intestine, they still exist in our respiratory tract. Here they help to expel slime laden with microbes and harmful substances out of the lungs and upper airways towards the throat. This is the 'mucociliary elevator'. Structurally speaking, the mucus layer in our airways calls to mind the image of toothpaste sitting on a toothbrush. The cilia themselves are nestled in an antimicrobial viscoelastic liquid which is covered by a double layer of gel. This is a striated organization like the mucus barrier in the colon – but turned upside down. The bottom layer of the respiratory barrier has the looser structure, while the tougher slime on top serves as a sticky trap for pathogens and harmful particles. The cilia have to move around freely in their only slightly viscous medium, but at the same time be able to reach the mucus barrier on top to carry it away. The removal process will come to a standstill if the gel is sitting too heavily on the cilia.

In many infections of the airways and in chronic conditions like cystic fibrosis, the stringy, sticky mucus associated with the illness cannot be adequately expelled. The result is an inflammatory response; molecules are released and attract harmful pathogens like *Pseudomonas aeruginosa*, which makes its home in the gel and can damage the tissue. Meanwhile, it has been suggested that the tough slime, produced in excessive quantities, combined with the ongoing inflammation may cause scarring in the lungs and contribute to an increasing loss of function as well.

Our expansive mucus barriers must perform a double function: keeping all pathogens away by trapping and outwitting them, while also playing host to a microbiota. Why are they not all built the same? They cover a wide variety of interfaces, with each location demanding a tailor-made defence, and they carry out other tasks as well. The gel layer on the surface of our eyes, for instance, is delicate and transparent so as to enable us to see through it. That is enough to act as a lubricant, helping the lid slide up each time we blink and not become stuck to the moist eye. At the other end of the spectrum there is the cervical mucus plug, which shrinks and swells, becoming more or less porous depending on the stage in a woman's cycle.

And our microbial friends? Their importance is never more evident than when they are absent. Without its microbiota, the polyp Hydra will usually die of a fungal infection, even if the creature is quickly being recolonized by some bacterial representatives from the old microbial cast. It seems that a few species alone cannot make up for the right mix of microbes, but which combination is the correct one? And does it even exist? From each holobiont's very beginning, no microbiota is set or static. The host's individual traits, its immune system, its slime barriers, but also its diet, any diseases, or contact with others all contribute to its internal population, making it unique and dynamic well into old age.

But general rules apply. Microbial variety in our internal ecosystems seems to make us more resilient. Real biodiversity, though, might only persist in pre-industrial societies, like an ark from another period of our evolutionary history as holobionts. We might not yet know in detail what a loss in microbial variety means to our health, but the modern lifestyle, with its unbalanced diet, excessive hygiene and microbe-killing medicines, has impoverished what was once a rich ecosystem. Stress and infections can also upset the balance. Dysbiosis is the term given to a systemic shift in the diversity and combination of species, and can play a part in many conditions, such as asthma and autism, auto-immune diseases and Type 2 diabetes.

Harmful bacteria like *Clostridium difficile*, which can cause diarrhoea, benefit from imbalances in the gut. Antibiotics are indispensable in helping us combat all kinds of bacterial pathogens, but they must be used with caution, because our own microbiota will suffer collateral damage. When the resident competition drops out or is subdued by antibiotics, pathogens are sometimes able to squat in the newly cleared-out gut. Another round of antibiotics might just let the vicious cycle repeat itself. Instead, science has developed a somewhat unorthodox approach – faecal transplants. Purified gut microbes from a healthy relative will be introduced to the patient's intestine, hopefully to settle there and kick-start a balanced microbiota before *Clostridium difficile* returns. It is a low-tech therapy with an extremely high success rate, and which might be able to help in cases of microbial dysbiosis in other parts of the body as well.

The balance of the microbiota is under threat right from the beginning. Many mothers in the natural world give their babies a basic microbial toolkit to start them on their way. Plants equip their seeds – and birds their eggs – with bacterial armour. The microbial conquest of the human child might begin in the womb before birth, according to a disputed publication. The child will then pick up microbes as it travels through the birth canal. A Caesarean section changes that cascade of colonization with bacteria from the mother's skin coming into play first and beating out those from the vagina and bowel. How much this affects the child's health is under debate. But it seems easy to rectify the disruption. In a recent study, a few infants born by C-section received faecal microbiota from their mothers and subsequently developed a regular microbiota.

Additionally, the mother's breast milk can help by providing more gut bacteria as well as mucins as decoys to trap pathogens. Once this microbial cord is cut, though, the child is obliged to cultivate its own microbiota that needs to carry out some basic functions first and foremost, be it by a core group of specific species or just any microbes that deliver. Microbial specialists will be involved as needed, depending primarily on the diet. In one study, the guts of American test subjects were shown to favour bacterial experts which specialized in breaking down fat and protein-rich food, while the microbiota of participants from rural Africa knew best how to crack the complex carbohydrates in beans and pulses. After just two weeks of swapping meals, the Americans were already exhibiting a slightly 'African' profile in their gut microbiota – as well as improved health parameters. A diet of chips, burgers and hotdogs saw participants from Malawi exhibiting 'American' characteristics in their blood tests for the first time, including slightly elevated markers for diabetes and bowel cancer.

Tireless and endlessly flexible, the microbiota serves the whole organism. But why aren't we capable of doing its job ourselves? According to one theory, early multicellular organisms may have been overtaxed by the increasing demands of their complex bodies. They may have outsourced certain digestion-related jobs to microbes, in return offering them a new habitat on the organism's surfaces, accommodation, protection and food. Not that they probably had a say in the matter. The first animals found themselves in a world which had been dominated by bacteria and other microbes for the previous three billion years or more. Microorganisms were central components as well as architects of the environment and they seized upon these multicellular newcomers as welcome habitats, firing the starting pistol which led to us developing into the organisms we are today.

IV
Life

As repellent as we find slime today, it has played a significant part in the history of science as the presumed link between the inanimate matter and life on Earth. And so, in the nineteenth century, the search began for a slime that was the source of all life, or at least the essence of life in material form. This was protoplasm, the slime of the cell, which was thought to receive and store signals from an invisible medium, the ether. They were thought to accumulate as highly individual patterns in the slimy substance, offering an elegant solution for the problem of heredity as well: could they be passed on from parents to offspring? This new and exciting research even trickled down into the world of modern art, including in the work of Edvard Munch. But what does evolutionary biology look like today? Somewhat surprisingly, slime is experiencing a renaissance, not as a giver of life but as a possible shelter for the first biological molecules and the precursor of early cells, which otherwise might not have survived in their extreme environments.

10

On the Hunt for Ur-Slime

The little mermaid lifted her clear bright eyes towards God's sun, and for the first time her eyes were wet with tears. On board the ship all was astir and lively again. She saw the Prince and his fair bride in search of her. Then they gazed sadly into the seething foam, as if they knew she had hurled herself into the waves.

Hans Christian Andersen, *The Little Mermaid*,
trans. Jean Hersholt

Are those the tears of a little mermaid I see caught in the waves lapping at my feet? They glitter in the water like diaphanous pearls, splitting the light into all its colours. I pick one up but it loses its shine. It becomes little more than a globule of jelly, only regaining its shimmer when returned to the water. Perhaps it's one of those seemingly alien comb jellies, after all, which resemble jellyfish but constitute their own taxonomic group, the *Ctenophora*. They are so old, they might just give sponges a run for their money as the first animal on Earth. The scientific debate over which group is the oldest has raged for decades, and it is still not settled once and for all.

There's a lot to be said for sponges in their simplicity, lacking any true tissue or organ and with their cells embedded within a jelly-like matrix. In contrast, gelatinous ctenophores are more complex in their construction, kitted out with bowels and a nervous system. Fringes of hair-like cilia, the comb rows, line the body and transport prey sticking to the ctenophore's mucous surface to the mouth. Their co-ordinated movement breaks light and lets rainbows of colour race along the length of their transparent bodies. It is mesmerizing, but does it make them the evolutionary pioneers?

Some publications have stated just that, only to be rebuffed soon after. To be fair, claiming that ctenophores are the oldest animals takes some serious explaining. Sponges are much simpler organisms and would have had to develop from more complex ancestors, losing already existing traits on the way. That is not impossible; sometimes evolution takes one or more steps back. But it is still more intuitive to let complexity arise from simplicity, and it seems that our evolutionary narrative doesn't have to tie itself into knots as sponges have a tentative hold on the crown – for now.

What this dispute does show, however, is how difficult evolutionary links can be to unravel, especially if they go back over hundreds of millions of years. As a consequence, we may never know the true shape of the tree of life – which looks more like a network of life anyway. Still, the tree has long been a metaphor for the histories of individual families. Royals and other powerful clans have recorded their ancestry in this way since the Middle Ages. And the botanical tradition has left its mark on language too: in German we still speak of the *Spross*, sprout or offspring; a first son and heir is a *Stammhalter*, literally 'upholder of the trunk', and 'branches' of the family may wither.

For many centuries, natural philosophers used the image of a tree to illustrate the interconnected nature of life, though this did mean that it became somewhat fixed. Any change, such as the extinction of a species, or its development, might have called into question the perfection of Creation – and thus the Creator. In accordance with this model, sponges, ctenophores and other animal groups were distributed across the branches of the tree of life like baubles on a Christmas tree, perhaps with some semblance of order but without a compelling context. However, this has fundamentally changed since Charles Darwin presented his theory of evolution to the world.

Darwin demonstrated that biological species emerged, evolved and even declined through a process of change and natural selection. This new vision of nature as an interconnected, ever-evolving system meant that we had to unpick the evolution of life over hundreds of millions of years. On the hunt for a suitable metaphor, Darwin tried and tested different ramified constructs, such as the river of life, the lungs of life and the coral of life. Ultimately, however, the classical tree of life prevailed, and the first traces of this can be found in Darwin's Notebook B.

It was on these pages that, in the summer of 1837, he sketched a small 'tree' – little more than a thin piece of scrub drawn at an angle – and noted down the words '*I think*' in the margin, presumably signifying his preference for this particular metaphor. This little tree is the only illustration in Darwin's *On the Origin of Species*, published in 1859. Here, though, the tree or plant has matured, growing dead-straight, its branches spanning the successive generations. It lacks any real details; for Darwin, it is a means of visualizing evolution. In *The Book of Trees*, Manuel Lima writes that Darwin was the first person to add changes over time, creating the first evolutionary tree of life.

Its precise configuration is something with which science has been grappling from the beginning. Thomas Henry Huxley called himself 'Darwin's Bulldog' for the grim determination with which he defended Darwin's then burgeoning theory of evolution, but he also threw his weight behind the ideas of other natural scientists when necessary. One notable relationship was with the naturalist, philosopher and artist Ernst Haeckel, himself an ardent proponent of evolutionary theory. If the theory of evolution had disproved the biblical story of Creation once and for all and it wasn't God who fired the starting shot of life, who let it first emerge from inanimate matter? Haeckel ventured an answer and declared primordial slime to be the source of life on Earth. That idea was not entirely new: from Ancient Egypt onwards, slime had been seen as a material between inanimate matter and life, as being able to bring forth at least some species.

In a sense it was only a matter of scale, with Haeckel's life-giving goo covering all or at least large expanses of the sea floor, relentlessly spawning new species. One problem remained, though: how do you catch primordial slime? Haeckel simply lacked the opportunities to prove its existence. As a naturalist he engaged intensely with various marine species, but a slime of the deep seas was beyond his reach and could not be captured by swinging a net from a rowing boat. It was here that technical advances and Huxley came to Haeckel's aid. Explorations of the deep ocean began in the mid-1800s to advance not necessarily science, but technology. The Atlantic Ocean floor was to be measured, its depth and profile recorded, in order to find suitable routes for undersea telegraph cables.

Messages between Europe and America had hitherto been subject to lengthy exchanges by ship. Why not use telegraphs? The technology required was already in existence. Samuel Morse had installed a connection between Baltimore and Washington DC, and cables had already been run along the English Channel. A transatlantic link would be much trickier to install, however, partly because the cable could not be borne by a single ship alone. After repeated failures and test runs the first communication occurred, between US President James Buchanan and Queen Victoria. But the connection faltered soon after.

It was to be an exciting time for the inhabitants of land – and water. 'Fishes and snails, everything that swims, everything that creeps or is driven by the currents, saw this fearful thing, this enormous unknown sea eel that all of a sudden had come from above,' wrote Hans Christian Andersen in 'The Great Sea Serpent', a kind of fairy tale:

> What kind of thing was it? Yes, we know! It was the great telegraph cable that people were laying between Europe and America. There was great fear and commotion among all the rightful inhabitants of the ocean where the cable was laid. The flying fish shot up above the surface as high as he could, and the blowfish sped off

> like a gunshot across the water, for it can do that; other fishes went to the bottom of the ocean with such haste that they reached it long before the telegraph cable was seen down there, and they scared both the codfish and the flounder, who lived peacefully at the bottom of the ocean and ate their neighbours. A couple of the starfish were so frightened that they turned their stomachs inside out, but in spite of that they lived, for they can do that. Many of the lobsters and crabs got out of their fine shells and had to leave their legs behind.

It would be many more years before a permanent and reliable connection was installed, though this was expected to endure indefinitely, if the hero of Jules Verne's novel *Twenty Thousand Leagues Under the Sea* is to be believed: 'The long serpent, covered with the remains of shells . . . was encrusted with a strong coating which served as a protection against all boring molluscs,' he writes in his fictional eyewitness account. 'It lay quietly sheltered from the motions of the sea . . . Doubtless this cable will last for a great length of time, for they find that the gutta-percha covering is improved by the sea water.'

The electrical sea serpent did not last forever, but the project was a great success nonetheless, made possible by shipping crews spending long years charting the ocean floor beforehand. HMS *Cyclops* took part in the initial survey, but one of its officers also sourced fresh material from the sea floor for an old friend with an interest in evolutionary biology: Thomas Huxley gave precise instructions on how to preserve the samples – strong alcohol seems to have had a starring role in the process. When Huxley began to examine the specimens years after the return of the HMS *Cyclops*, he was in for a surprise. The preserved seawater contained floating clumps of a gelatinous mass, similar to egg white, pitted with chalky fragments.

There was no doubt in Huxley's mind: this colourless, structureless matrix must be Haeckel's primordial slime, found alongside the remnants of its last meal. He named his discovery in tribute to its spiritual father. 'I have christened it *Bathybius haeckelii* and I hope that you will not be ashamed of your godchild,' Huxley wrote in 1868 to Haeckel. 'I am, of course, most especially delighted by *Bathybius haeckelii* and am very proud to be the godfather at its christening!' came Haeckel's prompt reply.

Other evolutionary biologists were beset by slime fever as well and would go on to try their luck on fishing expeditions. Different varieties of ur-slimes were brought up from the deep ocean, including some primitive variants found in Arctic waters. Since phylogenetic trees were in fashion anyway, they were immediately declared to be the even more ancient progenitor of all primordial slimes. Some explorers even claimed to detect signs of movement in some sort of *Bathybius*, probably even more so when observing the specimen aboard a rocking ship caught in a swell. Haeckel felt honoured and completely vindicated. Huxley wrote of the matter with zeal. Not everyone was totally convinced by the *Bathybius* phenomenon, though. The more cautious Charles Darwin, like several other natural scientists, remained rather sceptical.

Nonetheless, unexpected support for the idea emerged in Canada, not from the sea, but from the Rocky Mountains. Stone specimens had been discovered and interpreted as the oldest, earliest organisms ever recorded, and were bestowed with the most poetic of names. They were named *Eozoon canadense*, literally the 'dawn animal of Canada', because they were thought to have originated with first life on Earth. The similarities between *Eozoon* and *Bathybius* were deemed remarkable by some proponents. 'It casts a morning glimmer of understanding about the nature of ancient organisms and shows us the simplicity of life's functions and their substrates,' the German zoologist Alfred Brehm writes of the '*Morgenröthen-Thier*' (the dawn animal), but still hinting at its uncertain organic nature.

The dusty archives at the Hunterian Museum in Glasgow still contain a few pieces of *Eozoon*, which I was able to observe more closely under the microscope. I didn't manage to make out a dawn animal with my untrained eye, nor any hint of a primitive organism, just wonderful, delicate stripes in stone. But even the Canadian kinship couldn't save *Bathybius* in the end. The would-be animals encased in stone turned out to be mineral pseudo-fossils, and the notion of primordial slime was similarly short-lived and exposed as a mistake. The second half of the nineteenth century brought not only a new era of technology, with projects like the laying of the Atlantic telegraph cable; it also saw new disciplines with more rigorous methods established that the ur-slime couldn't withstand.

Oceanography, for example, began with the voyage of HMS *Challenger* from 1872 and 1876. In preparation, the cannon on board had to be dismantled and laboratories built. Only then would the crew be able to survey the oceans, taking sample after sample, studying myriad creatures from the unknown world beneath the waves. HMS *Challenger* was supposed to decipher the mystery of *Bathybius haeckelii* and reveal the secrets of its habitat. But samples from the sea floor came back empty.

Only once they had been preserved with alcohol according to Huxley's strict instructions did the jelly-like masses appear, swimming in the seawater inside. There was no doubt about it: primordial slime was an artefact of the preservation process; more precisely, it was calcium sulphate precipitated by the chemical reaction between alcohol and deep-sea mud. Sadly, Huxley's specimen of primordial slime has been lost to history. The aptly named 'Artificial *Bathybius*' brought up by the *Challenger* expedition is the faded sample I eventually saw in the Hunterian Museum.

Its collection is a cabinet of curiosities fit for a time when wishful thinking still went some way in explaining the natural world but eventually got crowded out by modern science. In the end, Huxley and Haeckel's primordial slime didn't mark the threshold between inanimate matter and life but the dawn of a scientific age. *Bathybius* was dreamt up to close a gap torn in the fabric of thought by Darwin's revolutionary theory of evolution, straddling romance and rigour even if ultimately they could not be reconciled. The death of *Bathybius* was a cruel blow to adherents of primordial slime. Huxley rowed back instantly, but Haeckel clung to his godchild for another ten years. His personal stakes were high: *Bathybius* was a cornerstone of his system of nature.

However, Haeckel was not alone in his ardour. In 1975, the science historian Philip Rehbock pursued the question as to why distinguished scientists were taken in by the illusion of primordial slime. He concluded that *Bathybius* provided a universal answer to many pressing questions in a number of disciplines. It explained the origins of life while presenting itself as the most primitive of all species. It seemed to be a source of food which would explain how life could exist in the frozen dark of the deep sea. It also seemed to provide evidence of protoplasm, the mysterious slime, which filled cells and supposedly represented life in material form.

11

The Origins of Life on Earth

King Solomon, the Queen of Sheba
And Hoover sprang from that Amoeba;
Columbus, Shakespeare, Darwin, Shelley
Derived from that same bit of jelly.
A. Guiterman, 'Ode to the Amoeba'

If a star has a cold, does its phlegm somehow find its way down to Earth? One day in spring, I notice milky clots of jelly scattered by a small pond on a patch of marshland. They are hard but elastic, the largest of them almost filling my palm. Centuries ago, finds such as these were thought to be phlegm from the heavens: *Sputum astrorum*, commonly known as 'star bogies' or 'meteor jelly'. The scientific explanation is more mundane. In spring, female amphibians bulge with thousands of eggs which are coated with jelly during egg-laying. If attacked by a fox or another predator, frogs and toads will often excrete premature spawn, which then expands in contact with dewy grass; it may also be regurgitated by the attacker.

We're safe from slimy star snot for now. But might Earth have caught another interstellar infection? According to some researchers, life originated elsewhere and spread across our planet as if it was just another place to settle. Most theories as to how terrestrial life began look for its source closer to home, though, to Earth itself. The planet took shape about 4.56 billion years ago, a glowing fireball of interstellar dust and gas. It was hellish: the Earth rotated in eight-hour days, generating massive storms, volcanoes pumped out toxic gases, while the young moon orbited so closely that it triggered powerful tidal waves in our primeval oceans.

This aeon is known as the Hadean after Hades, the underworld of Greek myth. It was during this time that, against all odds, life dared to take its first steps in a sheltered corner of the Earth. But how did it actually start? We have to travel backwards in time, climb down the tree of life and slip down the trunk. There are many branches to observe along the way, among them the birth of the eukaryotic cell, which is the first to boast a nucleus and, as an origin point of all higher organisms, ending the solo reign of the microbes. We might also encounter LUCA, the *Last Universal Common Ancestor* of all organisms on Earth. Charles Darwin brilliantly intuited that all modern-day species share a single progenitor.

But what did LUCA look like? Even modern science can deliver only a likeness by tracing the evolution of widespread and largely unchanged genes in recent species as the supposed descendants of the ancestor back to where they coincide. This point – not more than a microbial phantom – is LUCA, the sum of these artificially aged or perhaps rejuvenated characteristics. It doesn't necessarily represent the oldest microbe, of course; it represents the primordial ancestor out of which – thanks to genetic superiority, lack of competition, or sheer coincidence – life as we know it today emerged in its entirety.

LUCA is a theoretical construct, but there is plenty that its calculated characteristics can tell us. We know that our last common ancestor lived more than 3.5 billion years ago, deep on the ocean floor, tucked around hydrothermal vents pouring forth a mineral-rich solution which supplied the microbes with energy. LUCA is an already complex building block in our history, yet remains beyond our grasp. When we reach even further back, towards the deepest roots of the tree of life, the picture will become even more blurry. And if Ernst Haeckel's primordial slime *Bathybius* isn't the answer to the question of how and where life began, what is?

Reconstructing deep evolutionary history is so challenging because it occurred an inconceivably long time ago, on a planet which enjoyed few of the comfortable climes we know today. We find more questions than answers, but we can make some attempt at an explanation. What constitutes 'life'? Is it a single organic molecule or several working in unison? Must we look to a whole cell to cross the threshold into life? Science is divided on all of these questions, so it might be easier to agree on the conditions under which life is possible in the first place.

Reproduction is necessary to pass on one's own attributes to the next generation. Self-organization, a process which sees biological structures like organic molecules combine to create a unit greater than the sum of its individual parts, like the cell, is also essential. Another prerequisite is metabolism, which depends on energy. Then there's environment. The hellish conditions on Earth at the time make it hard to imagine that organized and co-ordinated processes such as these could take place without being disrupted if they lacked a barrier against the chaos of the outside world. But biological membranes that shield modern-day cells were the result of biological organization that could only develop while being protected. It is a chicken-and-egg-situation. So, how did early life on Earth take its first steps? When? And where?

The process began with prebiotic evolution, as chemistry made its first forays into biology, producing simple components which joined to make more complex structures: proteins and information carriers in the form of nucleic acids like RNA, and according to recent research maybe even its more famous cousin DNA as well. Together they were somehow able to form a simple cell. Hydrogels, as we will discover later, may have played a crucial part in this. We can also narrow down our estimates as to when this took place. Researchers in Canada have discovered four-billion-year-old rock which they believe bears traces of microbes: complex life. The dating is under debate, but other findings confirm that life might have emerged with surprising speed, developing rapidly – in geological terms – after Earth took shape.

The question as to where life began is also a tricky one. Could we all be aliens? Some astrobiologists shift the origins of life on our planet to outer space, suggesting that the molecules which kick-started life on Earth were imported by asteroid. This cosmic boost could explain why terrestrial life took hold so quickly despite adverse conditions. Asteroids have also been found to contain biomolecules such as amino acids, the building blocks of proteins. This does not answer the question of how life originated, however; it simply volleys it to an unknown and probably unknowable cosmic somewhere else.

Terrestrial scenarios are more often discussed, of course, including that of a surf zone in the ancient oceans or reservoirs of water in the ice, where early molecules might have been able to endure the low temperatures. It was actually Charles Darwin who introduced the notion of a life-giving pool in a letter to the botanist Joseph Dalton Hooker: 'But if (& oh what a big if) we could conceive in some warm little pond with all sorts of ammonia & phosphoric salts, light, heat, electricity et cetera present, that a protein compound was chemically formed . . .'. However, the great Darwin himself did not dare present his idea to the public. Perhaps responding to the question of the origins of life was indeed *ultra vires*, as he remarked wisely, that is, beyond the scientific possibilities of the time.

Yet the concept of the unhurried, comfortable origins of life persisted in the German poet Gottfried Benn's poem 'Gesänge' ('Songs') in 1913:

> Oh, that we might be our ancestors,
> A little clump of slime in a warm bog.
> Life and death, begetting and birthing,
> Would slip from our soundless secretions . . .

Here the poet wishes himself back among his ancestors, 'a little clump of slime in a warm bog', which, admittedly, probably speaks more of a sense of world-weariness than scientific analysis. However, after a lengthy lull, Darwin's pool of water as birthplace of life on

Earth has come into focus again, mostly in the shape of shallow, hot, volcanic salt lakes, since their cyclical drying-out, together with the sunlight, might have helped the production and stability of biomolecules. If correct, Darwin's prescience would make him a visionary, albeit a tight-lipped one.

Other theories state that life began around hydrothermal vents in the eternal blackness of the ocean depths. Black smokers, deep-sea hydrothermal vents, were discovered on the sea floor in 1977, producing streams of mineral-rich solutions at temperatures as high as 400°C. Dense beds of bacteria inhabit the vents' edges and form the foundation for ecosystems, including such prominent members as the giant, blood-red worms *Riftia pachyptila* in their white tubes. It was the first time evidence of life had been found in such extreme surroundings, devoid of sunlight and under high pressure. Since this habitat is somewhat reminiscent of how we envisage the Hadean era, black smokers were long favoured as the possible source of life on Earth, though they are probably too hot for fragile biomolecules.

However, brittle structures such as these could have fared better with alkaline white smokers, undersea vents up to sixty metres high, found in a region of the Atlantic known as the Lost City. They reach deep into the ocean floor where, according to one current theory, the first biomolecules formed, accumulating and interacting with each other. And perhaps competing, too. The vents are extremely porous, riddled with small niches and holes, where countless evolutionary test runs may have been carried out in parallel, like a stony incubator. It might have been a brainstorm of life, both sheltered from and maintaining contact with a destructive environment.

That's the theory, at least; finding the evidence to prove it is, as with all the other theories, near-impossible. And some ideas never really take off – at first. Since the 1940s, clay minerals have also been posited as a possible nursery for delicate, burgeoning life, because biological building blocks attach themselves to their porous surfaces, where they are protected and close enough to be able to interact.

The idea of a mineral shelter for the first organic molecules and processes has always been an outlier among the first-life scenarios, but got a bit more traction when scientists demonstrated that clay can take on a new shape under marine conditions – as a mineral slime.

In that form, clay may have been able to provide an early alternative to a complex cell membrane, offering much-needed protection even without a distinct barrier. In seawater, and probably also under the conditions present in the ancient oceans, these minerals are capable of forming a hydrogel to which biomolecules will happily attach themselves and accumulate. But this jelly is capable of more still: it protects nucleic acids like DNA, allowing them to carry out functions such as duplicating and transferring genes to proteins, and may even speed up productive processes while keeping destructive environmental factors at a distance.

According to this scenario, a clay hydrogel might have provided enough protection against the external environment for molecules and enzymes to create first life in the form of a protocell. Assuming this is true, why would it take clay to morph into a hydrogel? Under certain conditions nucleic acids like DNA and RNA, but also proteins, can themselves form hydrogels several millimetres in size. In other words, the cells' central biomolecules were able to form slimes on their own. Did hydrogels of any kind have a part to play in the development of the first cells? There can never be any definitive proof of this, but if we accept that recent genes point to long-lost ancestors like LUCA, then modern cells might retain a legacy of our slimy past.

Cells are not viscoelastic hydrogels per se, but they do resemble them in many respects. After all, a three-dimensional if ever-changing network of polymers – the cytoskeleton – binds liquid in the cells' interior. This creates a gel-like mass in which individual molecules and structures are still able to move freely. Organelles such as the cell nucleus and the mitochondria, however, seal themselves off with their own membranes and form demarcated departments inside the cell, each with their own unique access codes. Anything wanting to enter the cell nucleus, for example, must first interact with the highly selective hydrogel in one of the pores in its membrane.

The remaining molecules are distributed across the cell interior like an open-plan office. Chaos seemingly reigns, but molecular teams are able to conduct private meetings for specific tasks if necessary. In these temporary departments without borders, proteins or other molecules combine and condense to form so-called membrane-less organelles. The process is called phase separation and the result is often compared to a mixture of oil and water, both still liquid but separate nonetheless. But since a cell's interior is not liquid but rather viscous to begin with, might these membrane-less organelles be better described as gels in an already gel-like environment?

Either way, these temporary structures can control cellular reactions, support some and speed them up while preventing others. We have known about membrane-less organelles for little more than a decade, but they already look like a whole new level of regulation in the cell, involved in a myriad critical processes, from the expression of genes and production of proteins to stress responses. A quick and even more robust gelling might also create a plug for a wounded outer membrane and prevent the cell mass from spilling.

It is a fine balance for the cell to maintain between lightning-fast formation of these temporary structures and their equally quick dissolution. If these organelles become permanent or solidify, essential cellular processes might go haywire. And the consequences could even lead to serious illnesses if the disruptions become a pattern of frozen aggregates of molecule. Alzheimer's disease, for example, is an advancing neurodegeneration where insoluble clumps of protein accumulate in the brain. There are other forms of dementia and diseases like amyotrophic lateral sclerosis where this might play a role as well.

Deciphering these processes is slow work and it remains to be seen if any of the hypotheses hold true. It has been suggested, though, that bacteria – which lack any membraneous organelles – build condensates as well. Phase separation could very well be a universal phenomenon and might have played a role at the origin of life on Earth. What we do know is that hydrogels excel at keeping biomolecules in close contact, while stimulating their interactions and protecting them from a chaotic or hostile environment. Would it really be so far-fetched to assume that early organic structures and protocells were creating their own cosy gels as scaffolds, support and protection? More and more scientists seem to entertain the idea. But they are not the first ones: as early as the 1920s the Russian biologist Alexander Oparin proposed membrane-less jelly-like blobs as the origin of life.

12

The Substance of Life

> *They don't reproduce in their slug form as Terrestrial slime molds do. They seem to need the slug form only to produce enough of their corrosive antifreeze solution to enable them to migrate through rock to a fresh supply of food . . . And they ooze through the rocks in their slug form, their corrosive slime dissolving trails, cracks, and making more dust. These creatures are living Martians!*
>
> Octavia E. Butler, *Parable of the Talents*

Remember *Bathybius*? When children go off the rails, parents tend to turn on one another. A similar situation awaited the foster fathers of *Bathybius haeckelii*, that would-be primordial slime pulsating somewhere deep on the ocean floor. When the *Challenger* expedition revealed *Bathybius* to be an illusion, Thomas Henry Huxley was quick to offer up his mea culpa. He had christened his find and thought *Bathybius* would do him proud. 'But I am sorry to say, as time has gone on, he has not altogether verified the promise of his youth.' Huxley immediately apologized to everyone involved, but he was undoubtedly responsible, first and foremost, for the error.

Ernst Haeckel, however, continued to cling to his slimy fantasy for another decade and lashed out at all sides, filled with a blind love for his disappointing offspring. The *Challenger* expedition? It had only failed because *Bathybius* was not ubiquitous across the world. And Professor Huxley? He'd dealt the fatal blow. A true father wouldn't dismiss his child so easily, Haeckel griped. His stubbornness came as no surprise; *Bathybius* was both the deepest root and therefore the base of Haeckel's phylogenetic tree of life. But other, seemingly primitive organisms – little more than sophisticated droplets of slime – dwelt down there as well. 'He claimed to have observed single-celled organisms that lacked a nucleus, which he named *Monera*, and claimed that they were the most primitive and ancient type of organism,' writes the science historian Sherrie Lyons at the State University of New York.

So even if *Bathybius* was disproved, could not other primitive slimes still represent the key component of life at least? What looks like an obsession with slime on Haeckel's part drew on the notion of spontaneous generation, a theory with a long history going back to Ancient Egypt. It was supposed to explain why life forms seemed to spring suddenly from slime, mud and other products of putrefaction. Rotting meat suddenly teemed with maggots, and overripe fruits exploded with flies. The spontaneous generation of life seemed omnipresent. Only the cast changed over time: different kinds of pests were usually thought to spring from mud and slime, but sometimes higher organisms as well.

The idea of spontaneous generation owes its longevity at least in part to the prominent champion it found in Aristotle in the fourth century BCE. His work shaped science in Europe for centuries, but was itself frequently based on older concepts. Aristotle seized upon spontaneous generation, hoping to reinforce it scientifically. He studied nature, including the origins and reproduction of various organisms, which he categorized according to his own criteria. But not all species could be readily observed during mating or egg-laying. For Aristotle this meant that insects, snails, eels and similarly low creatures must emerge from dead matter.

In the thirteenth century, the friar and philosopher Albertus Magnus wrote of spontaneous generation as a function of the sun and the stars. According to him, their rays should be powerful enough to force the shapes of an entire bestiary on all kinds of materials. So strong was their all-pervading potency that they even gave stones the form of toads, worms and other creatures. There was a limit to star power, though, and these animals were never truly viable and saw out their existence as fossils.

The idea of spontaneous generation was still prominent in the eighteenth century and influenced even Jonathan Swift's satirical *Gulliver's Travels*, published in 1726. Swift's Houyhnhnms, a race of sophisticated horses, discuss whether the deformed and wicked Yahoos should be obliterated entirely from the face of the Earth, claiming that they 'privately suck the teats of the Houyhnhnms' cows, kill and devour their cats'. No one knew, though, how the Yahoos came into existence, whether 'by the heat of the sun upon corrupted mud and slime, or from the ooze and froth of the sea'.

The creatures' obscure extraction is probably intended to represent their lack of Christian beliefs rather than any genuinely humble origins, yet it does illustrate the popularity of the idea of spontaneous generation, both as a metaphor and as a scientific theory for the origins of at least some versions of life on Earth. For a time, it was even thought that human beings had been generated spontaneously, because how else could the Earth have been repopulated after the Great Flood? Usually, however, spontaneous generation was attributed to lower creatures; and by the nineteenth century it was only being suggested for microbes.

Still, that was an extraordinarily long life for an ultimately misguided theory. Spontaneous generation withstood the scientific revolutions at the beginning of the modern era and endured for millennia. 'That the idea of spontaneous generation survived careful experiment after careful experiment, radical thinker after radical thinker . . . is remarkable,' writes the author and scholar Daryn Lehoux, and mentions that it even survived beyond the invention of the internal combustion engine. 'Spontaneous generation was, in that sense, the last stand of the ancient scientific world view.'

But it wasn't just the emergence of species and their reproduction which caused confusion and gave rise to obscure theories. In the case of zoophytes it was their very nature which was disputed.

Zoophytes were thought to be plant-animal hybrids and evaded any clear classification. After all, weren't Venus flytraps, thanks to their carnivorous lifestyle, not at least as much animal as plant? The discussion surrounding zoophytes came to a head in the early eighteenth century as more and more mysterious creatures were brought to Europe – arriving, if not in physical form, in fantastical reports and stories. One such story told of the Scythian lamb, a sheep with a stalk. It grew from a supple stem and grazed the surrounding greenery, until it starved or was gobbled up by a wolf. Could this have been a reference to once-exotic cotton? Its German name, *Baumwolle* (tree wool), would suggest so.

Huxley marvelled at an even more peculiar life form, which, like fungus, lived on decaying plant matter but could transform into a creature capable of moving like an animal. 'Is this a plant or is it an animal? Is it both or is it neither?' Huxley asked his audience in a famous lecture on 'The Physical Basis of Life' in Edinburgh on 18 November 1868. 'Some decide in favour of the last supposition, and establish an intermediate kingdom, a sort of biological No Man's Land for all these questionable forms.' We now know that Huxley's hybrid life forms were likely slime moulds, which – and here's the source of the confusion – are not actually moulds at all. Nor are they fungi, as the German name *Schleimpilz* would suggest.

'Ni plante, ni animal, ni champignon, voici le blob!,' writes the French researcher Audrey Dussutour of these unique creatures in her book *Le Blob*. They are eukaryotic organisms that live solitary lives as single-cell amoebae, but under the right circumstances they can merge together to form a kind of multicellular creature – or a single cell which contains thousands of nuclei. There are two varieties. In leaner times, cellular slime moulds can gather in groups of up to two million amoebae, creating a slime matrix. Looking like a faceless, see-through slug just a few millimetres long, this so-called grex is capable of crawling and leaves a trail of slime in its wake. Yet the grex is only an intermediary stage, out of which a stalk with fruiting bodies ultimately develops, releasing the spores for a new generation of amoebae.

In contrast, plasmodial or true slime moulds fuse together thousands of cell nuclei in a single cytoplasm. This enormous sac of amoeba, technically still just one cell with one cytoplasm or inner mass surrounded by a sheath of slime, is called a plasmodium. *Physarum polycephalum* (the latter word meaning 'many-headed') is a favourite in the laboratory to demonstrate how this organism functions. And it shows a kind of intelligence, which proceeds with an almost mathematical precision despite the organism's lack of a brain, ganglions or neurons.

In *Le Blob*, Dussutour describes how she challenges her slime mould by setting up parcours with bait like rolled oats or repellents like caffeine. As it turns out, the social amoeba can learn: in the first few attempts it will crawl around – or rather, glide past – the bitter caffeine towards the oats. Later, the caffeine deterrent becomes increasingly ineffective, until it ceases to work altogether. The plasmodium has learnt to tolerate the caffeine, a new behaviour which it will even transfer to other slime moulds if it fuses with them.

These ruthlessly hungry slime creatures swap experiences and pass on information simply by merging. So simple, yet so beguiling and alien, they are reminiscent of Lovecraftian creatures. The science-fiction writer Octavia E. Butler (1947–2006) even researched them extensively for her works. Other fictional cousins of the slime moulds could be the amorphous 'symbiotes', ever-changing creatures from the Marvel universe of comics. These body parasites fuse with their helpless hosts, transferring to them their tremendous powers but also dark desires, yet remaining ever in control and pursuing their own agenda. Thanks to the 2018 film *Venom* these extraterrestrial organisms, capable of shape-shifting at will, were also immortalized on the big screen.

Terrestrial plasmodia inch forward rather more slowly by letting their cytoplasm stream in the direction of their choice and their cell membrane bulge out like long fingers. They can cover quite some ground – even square metres – as an ever-changing network of interlaced veins, branching out or pulling back as needed to explore other directions. But how does that work without a centre of command and with a flexible body plan? A plasmodium is a single cell with many nuclei, so should those 'heads' not be stuck in endless discussions and conflict about the right course? Maybe this works because the organism doesn't operate on a clear-cut yes-or-no basis but on shifting balances.

Within *Physarum polycephalum*, the slimy inner mass of cytoplasm flows in pulsing waves whose direction depends on stimuli, be they enticing or repellent. Light, for example, will chase the slime mould away in slow motion. But it is not an abrupt about-face, whatever that would look like in a slime mould. It is more like gentle ebbs and flows, each triggering locally a propagation of the signal – and the creeping waves gathering slo-mo steam. It is powerful collective yet sophisticated and rather complex behaviour that allows the organism to make economic choices, such as finding the shortest routes possible to food, while quickly cutting its losses on unproductive foraging quests.

Once a slime mould has decided against a particular direction or barren location, it will stay away for good, because a transparent slime remains, a ghostly echo of past routes. This serves as a reminder and warning for the slime moulds to avoid the area in the future and not spend any more resources in vain. Here we see gel used as an external spatial memory. Limpets use this technique too, following their own slime trails to find their way back to the safety of their homes before low tide in the chaos of the tidal zone. In all these and probably other cases, slime can even be used to communicate between members of the same species in order to find mating partners and – in the case of slime mould, at least – to keep competitors from approaching.

In another experiment to demonstrate the efficiency of the Blob's cool cost-benefit-analysis, Japanese scientists placed a plasmodium on the point marking Tokyo on a map while putting food on all the surrounding small towns. Soon afterwards, the shape-shifting slime had spread between the points using the shortest possible offshoots. Its slime network mirrored closely the actual rail network surrounding Tokyo, built over many years to connect the towns. Our own brains are networks of neurons with a definite structure. Other systems, including slime moulds, are changeable and function without fixed connections. In these cases, some scientists speak of a 'liquid brain'. *Physarum* functions a bit like a biological computer, making it of great interest to scientists working on artificial intelligence and new algorithms.

By now we know much more about slime moulds, but they are still tricky to classify. As Wikipedia has it, 'slime mould is an informal name given to several kinds of unrelated eukaryotic organisms.' Or are all these types of categories somewhat artificial anyway? In the same lecture in which Huxley lamented the ambivalence of the unusual hybrid nature of slime moulds, he also rejected the division between plants and animals as pure convention, because all life forms are the same on a fundamental level. But what constitutes life? Without the notions of God and the soul, all that is seemingly left is a purely mechanical interaction between cell components. But that wasn't enough for Huxley.

What he was looking for was that mysterious quality which distinguishes life from dead matter, and these reflections brought a long-discussed idea to the fore once again: life might be imparted by protoplasm, the slimy substance which makes up cells' interiors. One argument in its favour: protoplasm is shared by animals and plants alike. As Huxley's contemporaries could show, all cells were filled with 'a glutinous diaphanous substance that was insoluble in water and contracted into a spherical mass, sticking itself to the dissecting needles, that could be drawn out like slime,' as the American science historian Sherrie Lyons writes.

Some even 'maintained that the cell could be a complete morphological unit, even in the absence of a cell wall. Thus, many researchers thought that this protoplasm might be the "stuff" of life,' according to Lyons. This idea eventually led to the protoplasmic theory of life. In other words, it was the slimy mass in its interior that made a living cell. Protoplasm was said to be the substance and the basic unit of life. It was Huxley, however, who brought widespread and public attention to these ideas.

The smallest protoplasmic units were thought to be primitive life forms such as bacteria and amoebae. But they could also combine – a bit like Lego – to create larger organisms which, after their deaths, would disintegrate into their fundamental components for a new round of never-ending recycling. All these questions were considered, somewhat unexpectedly, in a novel by Thomas Mann (1875–1955). The German Nobel laureate, who frequently pondered medical and scientific theories in his work, has one character say in *The Confessions of Felix Krull, Confidence Man:*

> Oh, the idea is simple enough, just the cohabitation of cells, just the inspiration not to leave that slimy, glassy bit of primeval life, that elemental organism, by itself, but to construct, at first out of a few and then out of hundreds of millions, living designs of a higher order, multicellular creatures, great individuals – in short to create flesh and blood.

A system of easily recyclable basic units that could create complex organisms, so to speak. Early proponents of such a system included the German natural philosopher Lorenz Oken (1779–1851), who seemed to combine ideas of spontaneous generation with some sort of life-giving slimes – in a rather romantic approach. 'Every organic thing emerged from primordial slime and is little more than slime in various forms,' wrote Oken in his *Textbook of Natural Philosophy*, adding, 'All organic matter will dissolve back into slime, meaning that slime, having taken a form, will return to a state of formlessness.' But where does primordial slime, the source of all organic matter, come from? According to Oken, it is sea slime:

> This slime has its origins in, and is in its essence of, the sea, not mixed with it through the dissolution of rotting substances . . . All life is from the sea, not from the continent . . . All slime is living . . . The whole ocean is alive. It is an undulating, ever rising and ever sinking organism . . . Love sprang forth from the foam of the sea.

In England Oken was accused of heresy, because the notion of life having its basic origins in globules of slime from the sea still was seen to contradict the very foundations of theology. Yet the idea of living slime led to the golden era of protoplasm which Haeckel inserted into his system of nature in the form of *Monera*. After fishing tiny balls of slime out of the Mediterranean near Nice in the spring of 1864 that were probably amoeba-like organisms, he christened them *Protogenes primordialis*, 'firstborn of the primeval age'.

He saw them as the most minimal form of life, an idea that Huxley accepted, describing protoplasm as the potter's clay which, be it fired or painted, remained clay, fusing all life together. Eventually the protoplasmic theory of life had to be abandoned, not least because the nucleus and genetic material were understood as essential. But cell slime did not leave the stage without a fight – and not without leaving its mark in the arts.

13
The Art of Vibration

'No, sir. The ether. The earth browses upon a circular path in the fields of space, and as it moves the ether is continually pouring through it and providing its vitality.'
Arthur Conan Doyle, *When the World Screamed*

In the history of evolutionary biology, slime has been cast in many roles, including the source of all life and first life itself. Thanks to its seemingly unformed nature, almost any big question in science could surely be said to have an answer in slime. Haeckel's much-loved theory of primordial slime may have been killed off, but towards the end of the nineteenth century another biological slime enjoyed a surprising career as an evolutionary force: protoplasm, the cell's gently pulsating interior. It came to be seen as more than just an animated substance. Could it be the substance of life itself? And could it explain both heredity and memory as well?

According to this elegant new theory, neither cell membrane nor nucleus were needed as long as protoplasm acted as an animating force. Just like Haeckel's *Bathybius*, cellular slime turned out not to be the essence of life in the end. But it seemed so plausible at the time. It even sparked a short-lived craze in the sciences, in public life and in modern art. This 'protoplasmania', as the Canadian historian Robert Michael Brain dubbed it, lasted until the eve of the First World War. Brain's works have served as my primary references for this fascinating facet of slimy history.

In the nineteenth century, along with evolution, energy was the other big idea to define the sciences. At the time, physics sought to explain the propagation of light and electromagnetic waves. This seemed to necessitate the existence of some subtle medium, the ether. It is another notion that goes back at least to Aristotle, who described it as quintessence, the fifth element. What this mysterious medium was made of, whether it would be firm or gelatinous, no one could quite say. What seemed certain, however, was that thin strands of ether filled the entirety of space, vibrating and disseminating waves and light.

But what if vibrations were involved with the equally mysterious processes of life as well? After all, the slimy protoplasm inside cells was a sensitive and seemingly alive medium contracting and moving in waves. There seemed to be some parallels between the vibrating ether and pulsing cell slime, an observation that became a starting point for theories of protoplasm as the slimy essence of life and connection between the individual and the outside world.

Protoplasm was thought to pick up sensory perceptions, for example, from the outside world in the form of vibrations. But that couldn't be all: we learn and change in the course of our lifetimes, which means that our experiences must somehow leave their mark. New science and technology seemed to offer an explanation with devices like the phonograph, which the inventor Thomas Alva Edison unveiled in 1877. It is capable of recording and reproducing music and speech because a needle transfers the sound to a thin metal sheet in a pattern of waves. Consequently, it was thought that the soft cellular slime would retain echoes of sensory experiences and the like as physical impressions: they came in as ethereal vibrations and stayed as protoplasmic patterns.

With frequent repetition they would become permanent memories. It was fitting, then, that the prominent neurologist and psychologist Sigmund Freud thought of the contact points of neurons as protoplasmic threads. If memory could be neatly explained as durable impressions on cellular slime, could the puzzle of heredity also be solved? Another parallel seemed to present itself. In physics, a long and bitter dispute concerned the nature of light: whether it was a particle or a wave. In biology, however, the heredity concerned was the question of how the parent generation could transfer characteristics to its offspring – and again it hinged on the question of particle versus wave.

Charles Darwin speculated that the smallest parts of all types of cells in an organism, the particles so to speak, were collected in the reproductive organs. As *pars pro toto* parental samples they would be passed on to the offspring. Ernst Haeckel, however, advocated for the theory that every organism's traits were anchored and represented as ingrained waves in the protoplasm, a bit like grooves on a vinyl record. Every organism had its own unique pattern, combining characteristics it was born with and experiences in life that left their marks as well, as they were received as ethereal vibrations from the outside world. According to this theory, only protoplasmic wave patterns would be passed from sperm and ovum to offspring.

Particle or wave: who won? Light, as we know now, is both; interestingly, the same could be said about heredity, if not to the same degree. Our genetic material in the form of DNA is the primary and very physical store of information, with single genes as particles in a sense. They don't have to be transported to ovum as sperm, as Darwin thought, since every cell already contains the organism's complete construction plan. Genetic material, however, is also characterized by chemical modifications, which the parent and other progenitors have acquired over the course of their lives. This so-called epigenetics is an expression of someone's personal experiences and can influence offspring over generations. It doesn't work in the form of actual waves, though, protoplasmic or otherwise. Still, it allows us to visualize evolution as one long undulating wave, changing many times over and stabilizing intermittently as species and organisms, while standing for the fundamental kinship of all life. A bit like Haeckel did.

Where physics gave inspiration to many big questions in biology, another scientific discipline offered a more hands-on approach.

Experimental physiology is now an ancillary science, but at the turn of the previous century it was tremendously innovative and pioneering – even making the scientist as observer redundant. For the first time, machines were able to record different physiological processes over time, such as the heartbeat, visualizing them as curves, lines and waves. If the new field let diagrams and graphs speak for themselves, it wouldn't need human exchange any more and could bypass linguistic misunderstandings between scientists.

★★★

'All is fleeting; an insubstantial world, comprising only colours, contours and shades,' wrote the Viennese sensory physiologist and physicist Ernst Mach. 'Its reality is eternal motion, chameleon-like iridescence. It is in this game of phenomena that that which we call "I" crystallizes.'

Curves, lines and waves, colours, contours and shades, of course, had been established as a kind of wordless language for a long time – in the arts. But so far, the audience's reaction to a piece of art, be it music, literature or a painting, could hardly be predicted. Now it seemed that the new science of vibratory experiences could deliver the means to sharpen the artistic language into a tool. Might it be possible to target the audience's senses and sensibilities to trigger sensory impressions without the actual experience? Was this an opportunity to shape a new future for society, through art and the design of everyday objects?

In creating a kind of 'biological modernism', as Brain writes, science 'gave rise to a directly related quest among avant-garde artists for highly plastic, fluid, and mobile forms of art based in notions of rhythm, vibration, and periodicity, with a similar intention of altering the bodily sensorium of the beholder.' Among the more prominent proponents and practitioners of physiological art were the Norwegian painter Edvard Munch and the Swedish writer August Strindberg. They conducted their own scientific experiments but primarily tried to create a new art out of protoplasmic functions by being conduits themselves.

At the time artists had been seen as more intensely attuned than everyday people to the vibrations of the world through the eye, the ear and the nervous system. 'The great poet should not only perceive and distinguish more clearly than other men, the colours or sounds within the range of ordinary vision or hearing,' writes T.S. Eliot in an essay, 'he should perceive vibrations beyond the range of ordinary men, and be able to make men see and hear more at each end than they could ever see without his help.' Those exposed to this art and responding to it might perhaps unconsciously recognize a primal rhythm, something we possess as part of our protoplasmic inheritance.

The oneness of the world and its original nature must inevitably remain concealed from most humans, reliant as we are on our separate senses and only picking up snippets instead of appreciating it as a whole. How to surmount these divisions? Synaesthetes are people who see certain colours when they hear numbers or specific sounds. For protoplasmaniacs, the perfect embodiment of this ideal was the humble piddock *Pholas dactylus*, which uses a single organ to perceive light, sound, touch and smell in perfect synaesthesia. This would be an impossibility for average human beings but artists – as well as shamans – were thought at least to enjoy an increased sensitivity. Marc Chagall may have had an inkling of this: 'Colour is everything. When colour is right, form is right. Colour is everything, colour is vibration like music. Everything is vibration.'

Thanks to an artist's abilities, a walk while the sun sets blood-red might feel like the 'scream passing through nature' that Munch reported as his inspiration for various variants of his most famous work: *The Scream* blends the world's vibrations into shade and colour, shaping a distorted being in agony. Munch wanted to exorcize the experiences engraved upon his own sensorium on canvas. And if art is supposed to convey an artist's perception to the audience, this work has stood the test of time. '*The Scream* hits you like a bomb in black and white,' writes the British art critic Jonathan Jones in a review for the *Guardian*. 'The sky is a wobble of warped wood grain. Folds of black map the shore like ripples of trauma, crystallising in a lonely church tower. It's like looking at a heart monitor. The pulsations echo and amplify through space and you feel the same claustrophobic oppression that is tormenting Munch's universal figure of the modern soul.'

If vibrations such as light, sound, energy and radio waves constitute a world beyond human perception, and if any activity causes undulations in the elastic ether, this prompts the existential question: might the Beyond be trapped there as well? In the nineteenth century, new media such as photography, phonography and wireless telegraphy had proved that it was possible to capture and visualize vibrations, as Anthony Enns and Shelley Trower explain in *Vibratory Modernism*. Why not use new technology for supernatural phenomena as well? Occultism seemed to many people to be just another vibratory mystery for science to decipher by making the dead heard and seen again. Spiritual seances were all the rage.

And all things occult found an unlikely figurehead in the British doctor and author Sir Arthur Conan Doyle, famous for inventing the world's most rational detective, Sherlock Holmes. But later in life, Conan Doyle became a fanatical devotee of spiritualism, at least when not imagining the Earth as a somewhat protoplasmic creature invigorated by the ether. He took photographs at seances, capturing milky-white emissions emanating from the medium's various orifices. He declared that this 'viscous, gelatinous substance' was ectoplasm, related to protoplasm, and an expression of imperceptible vibrations. Countless props, from fluttering gauze to dolls and inflated gloves, can go a long way, especially if they can hardly be seen in dim light by an audience eager to believe.

It was perhaps a sign of the times that this particular fascination with the supernatural and its ectoplasmic embodiments emerged in conjunction with the re-emergence of the Gothic in literature.

As Kelly Hurley of the University of Colorado Boulder writes, the ruination of the human subject in often violent and repulsive terms is explored 'insistently, almost obsessively, in the pages of British Gothic fiction at the end of the nineteenth century and the beginning of the twentieth'. And what could illustrate the uncanniness of organic life more beautifully than goo? 'Slimy substances – excreta, sexual fluids, saliva, mucus – seep from the borders of the body, calling attention to the body's gross materiality,' writes Hurley.

Protoplasmania in science and in art came to an end with the First World War, though seances were more popular than ever before, with so many grief-stricken people seeking to hear their loved ones' voices once more. And yet, the use of fluid and viscous forms in art as a way of revealing the profundity below the surface of life has endured. Biomorphism is the style that tries to express the deeper nature of things, their swelling contours and irregular, organic shapes, preferring the curve to the poker-straight line. The critic Alfred H. Barr described a biomorphic piece by Hans Arp in 1936 as 'a kind of sculptural protoplasm' and spoke of 'the silhouette of an amoeba'.

Another legacy of this *fin-de-siècle* excitement about intangible influence is the idea that the public is in some sense a homogenous mass whose perceptions can be manipulated by art. Thanks to the advances in technology and data collection, it has become easy to find out when audiences are thrilled and when attention is lagging. E-readers make it possible to collect data which reveal the exact line where readers skip forward, for example, while online streaming platforms show audiences' reactions to films; whether the viewer watched to the end, paused the film or abandoned it. This minutely accurate information about our behaviour will be used to tailor future programmes to elicit the desired emotional response. Can culture be tuned to perfection? 'The user is a precalculated and predictable being, largely transparent in his or her behaviour,' writes the journalist Adrian Lobe in *Die Süddeutsche Zeitung*. 'Yet these calculations and forecasts of pleasure eliminate a variable which is fundamental to the business of culture and science: chance.'

The same chance makes some songs catch on and never let go – because they define a whole era. 'By 1966, the whole flower-power movement was well on its way,' writes the Beach Boys' Mike Love in his memoir. 'Hippies. Mystics. Longhairs. A new youth culture was rising, and rock and roll was its liturgy.' But one song would become a perfect hymn to this period of history, written by Love's bandmate Brian Wilson from a memory of a conversation with his mother. When he was a child, she had told Wilson that dogs could 'pick up' vibrations from humans. The word 'vibrations' had scared Wilson, Love writes. 'But it also appealed to him – the possibility of communicating on a non-verbal plane.' And the song? 'Good Vibrations', of course. *Oom bop bop.*

V
Evolution

In the beginning was the microbe. It conquered all the nooks and crannies on Earth and was coated in slime. It's possible that life has existed on Earth for nearly four billion years, and for most of history slime reigned supreme, a thick seal on the world. Until recently, modern science had thought this era exceptionally dull, a precursor to the moment when the development of higher organisms began in earnest. There is now speculation, however, that many cellular processes developed underneath this suffocating blanket of slime, enabling the subsequent explosion of more complex life. Early multicellular organisms emerged, such as jellyfish and other cnidarians, which were predominantly slimy and gelatinous.

14
City of Microbes

Chalmers, a man with a keen sense of humor, had used the example of a man looking down on a microscope slide and seeing the bacteria formed into the words 'Take us to your leader.' Everyone thought Chalmers's idea highly amusing.

Michael Crichton, *The Andromeda Strain*

He was particularly proud of his white teeth and took great pains over his oral hygiene, the biographer Laura Snyder writes of Dutch businessman and magistrate Antoni van Leeuwenhoek. In his leisure time, van Leeuwenhoek made glass lenses with equal parts relish and skill. Nevertheless, his academic contemporaries only began to take him seriously when they saw for themselves what his lenses made visible: the living, magnified *Animalcules* that we now know as microbes. Antoni van Leeuwenhoek had opened a window on to a new world and was now searching for traces of it everywhere – even on a September morning in 1683, as he was rinsing his mouth with saltwater, rubbing his teeth with a cloth and picking remnants of food from between his gnashers with the sharpened quill of a feather.

He was almost done when he found one last spot still harbouring a patch of plaque. He dissolved the goo in water and saliva and observed it under his glass lenses. He was surprised by the great number of minuscule creatures he could see. Some shot powerfully and speedily through the liquid 'like a pike', while others swirled around like 'gnats or flies'. His conclusion? 'There are more animals living in the scum on the teeth in a man's mouth than there are men in a whole kingdom', according to Snyder. It would take nearly 300 years before another man with an unusual hobby – this time mountaineering – proved that the image of the organized, structured kingdom was a surprising fit for the world of the microbes.

What does a scientist do if he wants to combine his work on bacteria with his passion for mountains? Well, he analyses the microbial contamination of alpine streams. At least, that was Bill Costerton's plan when he set off for a trip to the Canadian Rockies in the 1970s. Contrary to expectations, that natural bodies of water should be full of bacteria, the glacial stream proved to be almost germ-free. But the rocks on its banks were coated in a slimy biofilm which was so slippery that you could 'go right off your feet and land on your rear end in the water', as Costerton said in one interview.

It must have been a Eureka moment, as the biologist suddenly understood the significance of bacterial slimes. It changed the course of his career and of biomedical research, which had previously thought of bacteria as free-living unicellular organisms in a fluid medium and kept them that way in the lab. As the 'father of biofilm', Costerton established that this is, rather, the exception to the rule, whether in nature, in our bodies or in industrial plants. But what is a biofilm? Bacteria can live on their own in what is known as their planktonic form. Once they find a suitable surface in a moist or wet environment, though, they can attach. They're not picky: Costerton's riverbank, our teeth, medical implants in the body and an endless variety of other surfaces are equally attractive.

Other bacteria will join them there, be they from the same or other species or even a different kind of microbe. It doesn't matter if bacteria, archaea and fungi go it solo or mix and mingle – if they reach high enough numbers they will bunker down. In order to achieve this they produce an almost impregnable matrix, a slime that keeps stressors like mechanical forces or even antibiotics at bay. And the mature biofilm is a kind of hydrogel. Its matrix forms when microbes secrete what is known as EPS, extracellular polymeric substances or, in short, exopolymers. Once released, these highly diverse long-chain structures – with some sugary macro-molecules among them – form a gelatinous material composed of up to 97 per cent water. They give mechanical strength and structure to the biofilm, letting it resemble our own kind of cross-linked hydrogel.

What at first might look like a motley crew of microbes in a ramshackle slime shed is actually a complex community in a high-functioning and tailor-made environment with impressive defences. So impressive, in fact, that we are often quite helpless in our fight against microbial biofilms. Just consider Antoni van Leeuwenhoek's dental hygiene with rinsing, picking and rubbing. That sounds familiar, because it is: even today we have a hard time removing sticky microbial plaques from our teeth and are reduced to brute mechanical force by brushing and flossing.

Invincible microbial biofilms in our body can have far more serious consequences than cavities, though, and cause dangerous infections, as Bill Costerton was quick to recognize. Instead of continuing to focus on the spectre of bacterial baddies acting alone, he promoted the development of effective biomedical treatments against entire microbial communities and their goo. His findings are no longer disputed, but the task remains challenging. After all, biofilms are not only the oldest life forms on Earth; thanks to their extremely long evolutionary history, they are also universally successful and sophisticated masters of survival.

The origins of life on Earth may lie somewhere up to four billion years in the past, and the first life forms were probably microbes. 'I suspect that they were also capable of producing a kind of slime very quickly,' says the microbiologist Hans-Curt Flemming. Or there could already have been some slime present, he thinks. After all, protection in a dangerous environment is not only provided in sophisticated hydrogels of the biological variety that came later. Either way, since their early beginnings microbes have spread out in a tremendous diversity of nearly a trillion species, colonizing every tiny corner of the Earth and shaping it in their own interests.

Microbes are the engineers of all ecosystems, driving the planet's biogeochemical processes and cycles, breaking down and rebuilding matter of all kinds. The atmosphere was first enriched with oxygen only after cyanobacteria invented the process of photosynthesis. This particular kind of microorganism had learnt to use carbon dioxide (CO_2) and sunlight to produce sugar as a valuable resource, with oxygen as a by-product that is by now vital to us and many other organisms. Microbes are the foundation of the biosphere, our planet's biological shell, brimming with life. It extends kilometres down into the Earth's crust and up, beyond the clouds, into areas which are extremely hostile to life. The limits of the biosphere are those places where only the toughest microbes can at least temporarily tolerate the adverse conditions.

There are other hostile frontiers, like the interfaces between the spheres of air, earth and water. These are not only inhabited by microbes, they are shaped and governed by them too. Crucial to this success story are biofilms, the microbes' slimy protection against an aggressive environment and the foundation of their own co-operation. Microbes are capable of surviving without biofilms, but living as a community allows them to take on entirely new characteristics – a bit like multicellular organisms.

The earliest fossils of this kind of complex slime are at least 3.5 billion years old and originate from the Pilbara region of Australia. They are the remains of microbial stromatolites, which garlanded the oceans, forming the first reefs long before corals. Microbes embedded themselves in a robust layer of slime, a bit like the biofilms of today. Due to this microbial activity the matrix became enriched with minerals and unusable as a home. So now it was time to build upwards. A new layer of slime was added till in turn it mineralized as well. Thus, stromatolites grew layer by layer like a stony pile of pancakes. Their slimy history can still be read in the laminated appearance of fossilized stromatolites.

When did microbes learn that they fare better as a team in a house of slime? It must have been early on, even if we will never know exactly when and where this took place. But it is indisputable that biofilms are the first, the most widespread and the most successful life forms on Earth. And, in a sense, the slime matrix is their biological dark matter: 'So many questions remain unanswered,' says Hans-Curt Flemming. 'Slime provides a perfect basis for life inside the biofilm, but we're still unable to say which microorganisms form which components at what time.'

What is certain, however, is that the multifaceted and adaptable matrix protects its inhabitants and keeps them together, making life possible for them under extreme conditions such as those found in hot springs, in icebergs, in radioactive water or under high pressure on the sea floor. The matrix is not an inert jelly, it is a highly organized city of microbes – even if they're not yet communicating with us, as Michael Crichton suggests. So how is it built? First of all, microbes must gather in sufficient numbers. They arrange this via chemical signals, allowing them to estimate how many are present.

When their numbers are great enough they form a biofilm. In cases of disease-causing pathogens in human beings, it is often at this point that symptoms of infection begin to present, when the microbes are well protected and much harder to fight. How can we prevent biofilms from forming in the first place? Scientists are trying to understand microbial chatter better. If this communication could be interrupted or manipulated, even dangerous pathogens might be suspended in harmless solitude. One of the mucins in our saliva already seems to excel in this manoeuvre by keeping some caries bacteria in a kind of quarantine. The microbes survive, but they do not congregate or build a slime house.

If a biofilm does form, however, this matrix is the microbial community's interface with its environment, an intermediary for internal processes and interactions with the outside world. It is much more than an amorphous gel; it's a miracle of versatility. Its construction is not uniform but can make certain areas rigid, while others remain more viscous for mobile microbes. Locally specific differences in pH, temperature and oxygen content allow it to shape spatial architecture, complete with pores and different habitats. There's a microclimate to suit every need and preference.

Even diverse communities can find accommodation here and work in a way that suits them. Needs change, and the matrix is subject to a constant, dynamic process of reconstruction. This takes a lot of energy but it appears to be worth it, because, for one thing, communal processes are more efficient and increase chances of survival. For example, microbes exchange genetic material in the matrix and use it in a concerted way to help the community in big tasks like fighting antimicrobial agents, almost as a multicellular organism would handle defence, nutrition and waste removal.

When it comes to protection, the matrix itself is hard to penetrate, both by medical drugs and by dangerous attackers like bacteriophages. These viruses exclusively target bacteria which are often safe while hidden in the biofilm. It seems possible that they can even weaponize their own attackers. If phages get trapped on a biofilm's sticky surface, they might remain there as a layer of protection against unwanted bacterial invaders trying to gain entry into the matrix. That leaves more time for the resident microbes to search for food.

It's another communal task: nutrients are absorbed by the sponge-like surface of the matrix or detached from the underlying substrate. Wherever they come from, they are shared and distributed across the matrix via channels and transported so deeply inside the biofilm that every microcolony is fed. When microbes die off, they are taken apart and put to use by the rest of the community. Where there is food distributed, there is waste to be removed. Unwanted materials are gathered in one place, to be broken down by collectively secreted enzymes. This division of labour amounts to a communal waste disposal system which functions via external digestion.

Taken together, biofilms are high-functioning cities for microbes where everyone gets their fair share. The matrix can even go so far as to distribute sunlight evenly throughout the biofilm, so that everyone benefits. Still, all microbes are free to leave. Those who stay enjoy a choice of lifestyles: some microbes prefer to be loners, while others join up with members of the same species or get together in a more diverse colony. The community offers life on several different levels, but it's more than a city of microbes – it's an entire ecosystem.

15
One Long Melody

If the history of life were crammed into a year, slime would have ruled through spring, summer and autumn, continuing well past Halloween into the beginnings of winter.

Gabrielle Walker, *Snowball Earth*

Composed by the American musician John Cage, *ORGAN2/ASLSP* is an organ piece currently being performed at St Burchardi Church in Halberstadt, Germany. It is to be played ASLSP – 'as slow as possible'. It is intended to play for 639 years, the approximate lifespan of an organ, making it the longest and slowest piece of music of all time. It might be impossible to conceive of the scale of this composition, but it could help us to understand the history of slime, which is thought to have ruled the globe solo for at least three billion years. This length of time is even harder to conceive, beyond all human measure, and impressive even on a geological scale.

The piece performed in Halberstadt began on 5 September 2001, with a seventeen-month rest. There might have been a similar prelude on Earth, spanning the few hundred million years following the planet's birth around 4.6 billion years ago. It was only then that life emerged and likely took its first steps in the form of microbes, which might have left jelly-like tracks in their wake for the first time. It's conceivable that soon afterwards slime went beyond these humble beginnings and launched into its big solo, continuing, without stopping or losing steam, for the next three billion years.

John Cage's composition seems a little fidgety by comparison: it changes nearly every few months or at least years. One sound is silenced, another is added; the piece is currently stagnating at C, D flat, D sharp, A sharp, E and the more recently added G sharp and E, a seven-note chord that will be playing till 5 February 2022. In contrast, slime's theme has been playing for almost all of Earth's history and with only minor changes for most of that time. Its tune can still be heard today; it may even be the dominant note in places, but it is no longer a solo. The rise of the macroorganisms changed that. First there were soft-bodied and mysterious creatures, possibly somewhat gooey themselves. And then, with start of the Cambrian Period around half a billion years ago, other voices emerged. Today, we are experiencing the polyphony of the biosphere.

But back to the unimaginably long time from early microbes to the first animals when slime reigned unchallenged – and stifled the globe. 'Our home planet spent most of its long history coated in nothing but simple, primordial slime,' Gabrielle Walker writes of this phase of evolution. Do we struggle to imagine such vast expanses of time, or do we find monotony depressing because we crave change? During a visit to the church in Halberstadt a few years back, I found the much simpler but equally enduring chord fairly irritating; my ears even began to pick up changes which weren't really there. Our lack of patience is as good an explanation as any for the astounding fact that we labelled the last and most intense phase of slime's long reign, an era from at least 1.8 billion to 800 million years ago, 'the Boring Billion years', if not 'the dullest time in Earth's history'.

The Boring Billion? That needs a bit of unravelling. As stated before, from fairly early on in our planet's existence microbes were not only the first but the only life form on Earth and they were living in slime. Small in stature but big in consequence, microbes shaped the planet's geochemical processes and were catalysts for change on a global scale. Cyanobacteria, for example, learnt multiple times how to produce oxygen which filled the oceans and atmosphere. But then things changed. Around 1.8 billion years ago, oxygen contents in the air and water dropped dramatically to about 10 per cent of what we have today, and nutrients such as nitrogen became scarce. For a billion years the Earth was very probably climactically and geochemically stable, which is bad for life, because stability usually means stagnation.

The Not So Boring Billion? According to the fossil record, evolution appears to have stopped – or so the conventional thinking went. But there was at least one extremely exciting development in the emergence of the eukaryotic cell as a competitor to microbes. It seems that evolution was limited less by ideas than by a lack of materials and fuel, which became scarce during that time. And this worked in slime's favour. The ocean floor was covered in microbial mats as far as the deep sea. Much like the thinner biofilms, they contain sticky exopolymers, produced and released by microbes. Unlike biofilms, however, microbial mats can cover many kilometres in a layer which is tough and compact enough to be lifted off in one piece. Anthony Martin at Emory University compares it to cellophane: the sea floor was completely sealed and cut off from the seawater by the stifling blanket of slime.

The Exciting Billion? We now have an alternative reading of the seemingly monotonous era of slime: as a slingshot for complex life. It suggests that evolution was not stuck under a blanket of suffocating slime but rather found a cosy hiding hole there to come up with innovations like multicellularity and sexual reproduction. That kind of radical progress was born of necessity: the lack of nutrients may have sent the evolutionary engine into overdrive, at least in the first half of the Boring Billion, which then gave way to a phase of diversification, as nutrients were no longer in such short supply. Fundamental cellular processes and genetic blueprints may have been sketched out at this time, contributing in the subsequent Cambrian Period to an enormous, rapid development of complex species. The foundations for the Cambrian explosion of life were built under slime.

Before the Cambrian, however, there was yet another crisis, and a first, cautious flourishing. Some scientists speculate that the planet experienced a period of tremendous cold prior to the Cambrian, in which the Earth was almost completely covered in ice for around 200 million years. It was the coldest, most dramatic and greatest shock that the Earth had ever experienced, as Gabrielle Walker writes: 'It was the worst catastrophe in history.' According to this scenario, the 'Snowball Earth' had barely thawed when the Ediacaran Period began around 635 million years ago, during which mysterious beings appeared on the scene. They resembled metre-long feathers, discs with gills and jelly-like blobs and probably populated the dark and peaceful depths of the ocean, which were still covered in rubbery microbial mats at the time.

We can only speculate that these simply constructed life forms were anchored to the sea floor and took in nutrients from the water directly via the surfaces of their gelatinous bodies. Were they even animals? We still know too little about these creatures, which are sometimes considered to be unicellular organisms. Recently, though, there are some suggestions that they might have been more complex than expected; they might even have been the first animals after all. Still, biologists keep their cards close to their chests and prefer to talk about the Ediacaran biota than any real fauna. For the time being, though, it is still accepted that their era ended as abruptly as it had begun, like a failed evolutionary experiment. And so with the Cambrian era began the modern age of life.

An understanding of the astounding biological innovations of the Cambrian Period – which began around 540 million years ago – only began to emerge at the start of the last century. One December day in 1909, the American geologist Charles Doolittle Walcott discovered an unfamiliar fossil in the Burgess Shale of the Canadian Rocky Mountains. The creature, just a few centimetres in size and preserved in minute detail, distantly resembled a crab and it was not to be the only palaeontological curiosity found there. In the following decades, Walcott collected around 65,000 fossils from the same site. The bizarre creatures had one thing in common: they had all lived during the Cambrian Period and witnessed a biological Big Bang which is thought to have revolutionized life on Earth in a comparatively short period of time.

According to more recent findings, what is known as the Cambrian explosion may well have unfolded less suddenly in a few steps and even begun as early as in the Ediacaran Period, overlapping in part with its characteristic biota. But these are nuances alongside the established fact that during this time highly complex multicellular organisms first developed – and in incredible variety. Walcott's fossils are highly detailed, yet at first it was hard to assign to these creatures any relatives among recent species. Much later, though, most of them could be confirmed as early representatives of the animal groups we know today.

They brought with them breathtakingly modern blueprints and novel anatomical extras. Segments! Legs! Eyes! The world had never seen the like before. The animals sported equally innovative structural defences, with spikes on their backs and hard shells. Rather than being a decorative luxury, this armour may well have been born out of necessity, as carnivores emerged for the first time and helpless prey animals were forced to adapt. The biggest predator was *Anomalocaris*. This translates as 'abnormal shrimp', a fitting description of the creature, which was up to one metre long and a highly efficient swimmer. It roamed the Cambrian seas like a wolf in the fairy-tale forest, always on the lookout for prey with its powerful compound eye ('The better to see you with'), its spiny and mobile appendages ('The better to embrace you with') and its round mouth ('The better to eat you with'). The peaceful Garden of Ediacara was no more.

Those who were not keen to fall prey to the first carnivores had to be speedy, rely on camouflage or invest in defence. It's possible to view the evolutionary progress witnessed during the Cambrian Period as an arms race between the species. New interactions between predators and prey changed animals and their habitats, going deeper under the sea, and much later seeking out dry land.

In contrast, the Ediacaran biota was still a two-dimensional world a bit like Edwin A. Abbott's description of it in his novel *Flatland*:

> I call our world Flatland, not because we call it so, but to make its nature clearer to you, my happy readers, who are privileged to live in Space. Imagine a vast sheet of paper on which straight Lines, Triangles, Squares, Pentagons, Hexagons, and other figures, instead of remaining fixed in their places, move freely about, on or in the surface, but without the power of rising above or sinking below it . . .

The Ediacaran creatures were overwhelmingly stationary, fixed to the ground by gelatinous sheets or surrendering more or less passively to the current. Meanwhile, microbes were still biological master builders, covering and sealing these surfaces with their robust mats, just as they had since the dawn of time. But not for much longer.

★★★

During the Cambrian, complex life forms appear to have been rejigging their habitats with similarly complex lifestyles. Older traces remain of creatures mostly creeping and crawling on the microbial mats, probably to take in organic waste matter. Now, however, some species began burying themselves in and underneath the layer of slime, or even drilling their way into the sediment, perhaps because they were in flight mode or searching for new resources and prey. This had global consequences because the compact microbial mats were replaced by soft sediment once their slimy seal was broken. For four billion years the sediment had been home to microbes, for whom oxygen is toxic, and which now found its way into the sedimentary depths via seawater – and probably caused a mass die-off.

Nutrients were now being exchanged between sediment and

ocean. Across the globe, elements like carbon and nitrogen were being cycled between these two formerly separated spheres. These and other dramatic changes revolutionized marine habitats and shaped the evolution of animals. A shift in behaviour can be witnessed in trace fossils like burrows, borings and tracks, the focus of ichnological research. Fossil marks from that period also suggest that the Cambrian didn't come out of nowhere. After all, mobile animals were starting to dig into biomats and eating them as early as the Ediacaran Period. They must have developed suitable tools, at least for scraping if not breaking into the slime layers. Nevertheless, it was during the Cambrian that animals first became engineers of their own ecosystems.

Habitats became three-dimensional: digging brought new resources and refuges, as prey animals were able to disappear underground and predators were able to hunt them there. The late German ichnologist and expert of the Ediacaran Period, Adolf Seilacher, spoke of the 'Cambrian Substrate Revolution', when stifling microbial mats were bulldozed for the first time and had to retreat to areas where less bioturbation occurred, while churned-up sediments helped establish new habitats at the interface of formerly separated spheres. Anthony Martin, who specializes in ichnology, stresses that this dramatic change made the biosphere, geosphere, hydrosphere and atmosphere parts of one unified global system.

Slime's unchallenged billion-years-long reign is long over. But microbes in their slimy covers still shape global processes, and even robust microbial mats cling on in some remote corners of the Earth. And they could stage a comeback. The worst extinction event in life's history occurred 252 million years ago in the Permian Period and led to the staggering loss of nearly 96 per cent of all species. Many digging and grazing species died out as well, which might have helped dense microbial mats to come back. This does not bode well for our own time, when the so-called sixth extinction might leave ecosystems equally vulnerable. Will the loss of marine species let slime suffocate entire habitats on the ocean floor once more?

16

The Gelata Era

From the deck of a vessel you may look down, hour after hour, on the shimmering discs of jellyfish, their gently pulsating bells dotting the surface as far as you can see.

Rachel Carson, *The Sea Around Us*

At daybreak on 3 November 1944, huge balloons rose slowly into the air above a sparsely populated patch on the coast of Japan. Each was fitted with an incendiary bomb. It had been intended that the local populace be kept in the dark about the military initiative, but the fleet of airborne ships could not be concealed. One astonished observer is said to have compared the balloons, still to be fully inflated and dragging long cords along behind them, to gigantic jellyfish floating across the pale sky. It was an elegant image, yet this fleet was carrying Japan's lethal vengeance across the Pacific.

During the Second World War, Japan was hopelessly outnumbered by its enemy, the US, and when the main islands were attacked by US bombers, the Emperor of Japan demanded a new weapon to ensure victory. The plan was to shower America with bombs from the sky, but the distance was too great for the Japanese aeroplanes. It was at this moment that an obscure scientific publication emerged, telling of winds which blew in a powerful surge at altitudes of around 10,000 metres all the way from Japan to America. Jet streams such as these had been revered as *kamikaze* or 'divine winds' since the thirteenth century, when they had prevented the Mongolian fleets from invading Japan.

So why not entrust a new intercontinental weapon to these seemingly sympathetic currents of air? The *Fu-Go* balloons were ten metres in diameter and took two years to build out of light but tear-proof paper, handcrafted from the bark of the *Kōzo* (mulberry) tree and sewn together by schoolgirls who, despite having no idea of the nature of the project, were chosen for their delicate fingers. More than 9,000 balloons were eventually sent on their way, filled with hydrogen and loaded with incendiary bombs, ready to terrify the Americans.

Emperor Hirohito, a divisive man in every respect, had called for action. When I think of him the image of gigantic jellyfish is never far from my mind, so deadly and yet so elegant, as ambivalent as the *Tennō* himself. On the one hand, he remains to this day a controversial warlord. But was Hirohito a quixotic ruler, little more than his generals' stooge, or was this image fabricated by American occupying forces after the war, because he was of greater use to them as an imperial puppet than as a defendant before the court martial? Whatever his posthumous reputation, the role of supreme commander contrasted starkly with the Hirohito who had been fascinated with oceanography since his youth, who went fishing every week in Sagami Bay south of Tokyo for sea creatures to examine in his laboratory in the imperial palace.

Hirohito's great love were hydrozoans, a large class of jelly-like cnidarians that includes both Hydra, a harmless freshwater polyp, and the Portuguese man o'war, a siphonophore or colonial organism composed of around 1,000 differently specialized polyps. Some form highly poisonous tentacles, while another one creates the gas-filled bladder which *Physalia physalis* uses to float and sail on the surface of the ocean. It is a thing of beauty and danger, much like a *Fu-Go*. It may have been unusual for a Japanese emperor to study sea creatures, but Hirohito was just one of several monarchs devoted to science.

Prince Albert I of Monaco (1846–1922) spent many years travelling the oceans as far as the Antarctic, researching marine biology, an interest which he popularized in Europe and America, at least among sufficiently well-off amateurs. He later had a museum of oceanography built so close to the sea in Monaco that the building seems to sprout from the Monégasque cliffs. Inside, there are art nouveau chandeliers inspired by the elegant lines of Ernst Haeckel's illustrations of delicate sea creatures. His *Art Forms in Nature* also inspired a design on an even greater scale in the gateway the architect René Binet created for the 1900 World Exhibition in Paris: visitors entered through a giant metal structure with intricate openwork, just like the filigree calcareous skeleton of the single-cell radiolarian on which it was modelled.

These days, there is some suggestion that both sides inspired one another, that Haeckel himself was influenced by art nouveau, given the flawless beauty and perfect symmetry of his illustrations. Yet Haeckel's aesthetic aimed not only to represent the beauty of these mysterious beings, to inspire artists and the public alike, but most of all to gather scientific knowledge from them. Since they inhabited an early stage of animal evolution, Haeckel hoped that simple gelatinous creatures like siphonophores, ctenophores and tunicates, but also worms and molluscs, would help us understand the origin of life and answer some of the basic questions of evolution.

The marine biologist Steven Haddock calls this the first Golden Age of the *Gelata*, a term he came up with for gelatinous creatures in the ocean. Never have so many hydrozoans and comb jellies been identified in so short a time as in the first decade of the last century, he writes. The scientific interest in these organisms was overwhelming but hard to satisfy: they are infinitely fragile and their delicate bodies fade and shrink in preservation, which makes them hard to study. And Haeckel's illustrations – idealized or not – only pinned them in two dimensions on the page. The drawings were beautiful, but not more than a reminder of the real animals' intricate complexity.

Two men found an ingenious way of catching the soft-bodied beauty of the marine world in astounding verisimilitude. Leopold and Rudolf Blaschka, a father and son from Bohemia, were glass-blowers. While on board a ship travelling to the US in 1853, Leopold was captivated by wonderful creatures in the water – just like the American author and marine biologist Rachel Carson would be much later. In the decades to follow, Blaschka and his son would create thousands of glass models of mostly marine invertebrates in their Dresden workshop for museums and universities. Later, the Blaschkas turned to plants and created a glass herbarium for the Harvard Museum of Natural History. The pieces can still be seen today, displayed in flat glass cabinets in their own room, labelled like botanical exhibits and absolutely lifelike.

Unfortunately, just like living things, the Blaschka invertebrates are now decaying; many have been forgotten and are damaged, are lacklustre and dull, if not coming apart or breaking. A renewed interest in these pieces might help; there are efforts to correctly label them after more than a century, to restore their former beauty and repair them. That leaves but one crucial question: how did the Blaschkas achieve such finish, colours and radiance? No one really knows how they worked the glass to craft these unique portraits by hand. But why did the beautiful pieces fall into disregard in the first place? It was not the Blaschka models the scientific world fell out of love with – it was the *Gelata*.

As Haddock writes, technological progress in the twentieth century eventually brought an end to the boom which had gripped the world of science. Now, instead of focusing on individual specimens of these delicate sea creatures, heavy machinery was used to collect them, and as many as possible. Ever-swifter ships would string out large nets on winches. The delicate *Gelata* did not survive the rough handling and there would often be little left of them but sticky slime. Gelatinous creatures were fading into obscurity too, as science concentrated increasingly on global oceanic phenomena and *Gelata* were only considered relevant en masse, if at all. Less attention was given to individual species and their lifestyles.

But this may be changing now with a new Golden Era of the

Gelata on the horizon. Studies show that gelatinous animals in the ocean are essential to marine food webs, both as predators and as prey. Haddock himself has played a part in a study which shows that siphonophores and ctenophores are important predators in some marine habitats. Jellyfish, on the other hand, are hunted themselves, by great ocean sunfish and leatherback turtles, which shred their prey with sharp appendages in their mouths and jaws. These few *Gelata*-munchers were long considered a curious anomaly, because their prey offered so little in terms of energy.

Why hunt a creature for little more than a mouthful of slime? We now know, however, that other predators, including some fish, penguins and albatrosses, are in fact 'gelativores'. *Gelata* are a central element of marine food webs, but how do predators use them to improve their energy intake? One suggestion is that they concentrate on energy-rich parts such as the reproductive organs or the stomach, which is filled with other prey. It is likely that they primarily eat easy-to-find *Gelata*, which do not require lengthy periods of searching and hunting.

And there's more: as part of the global carbon cycle a biological pump transfers bound carbon from the surface layers of the ocean into the deep sea, as we will later see in more detail. Sinking waste products, as well as dead animal bodies, make up a big part of that carbon cargo. Slimy substances come into play because they glue tiny organic particles together, making up 'flakes' that are big enough to sink as 'marine snow' into the deep. Recently, studies have shown that gelatinous zooplankton – jellyfish, ctenophores and free-living tunicates – are a major factor as well: these animals pack a lot of organic carbon into their bodies and dispose of parts of it in a way that will make it sink rapidly, just like many of their cadavers will do eventually. The yearly amount of carbon that ends up in the deep due to these animals alone could rival the amount the European Union releases into the atmosphere in the form of greenhouse gases – which should make *Gelata* a focus in climate-related research.

From a scientific point of view, so much about these animals is still unexplored and unknown. Like jellyfish, ctenophores and tunicates, many *Gelata* have complex life cycles, the details of which we have yet to uncover. These ancient survivors could be important in more ways than one when it comes to climate change. Oceans are undergoing change on a fundamental scale, becoming warmer and more acidic. This might be shifting the balance of species, possibly in favour of *Gelata*. We risk a new 'slime surge' if our soon-to-be empty oceans are home only to jellyfish, algae and microbial mats. This is why it is so important to solve the mysteries of the *Gelata* and understand their influence on marine habitats, arming ourselves for the future.

We need a new Golden Age of *Gelata*, and we might just have the means to achieve it: many of the new findings involving gelatinous sea creatures have been made thanks to technological innovations such as manned submarines and cameras which can be attached to other sea creatures. *Gelata* are now mostly observed in their natural habitat or carefully captured in containers, so that they can be examined on board or later in the lab. There has also been progress elsewhere. New technologies rely on soft robots that are made of flexible materials such as hydrogels. They are often invisible and capable of carrying out delicate work. There already exists a transparent gripping arm which works underwater and can grab fish or sea cucumbers and even jellyfish without harming them, making the individual animal the centre of attention again.

But what happened to those jellyfish-like wind ships that Emperor Hirohito launched in the sky? Several hundred went down on the east coast of America between Alaska and Mexico. From a military perspective the campaign was a failure, because the damp forests there did not catch fire. Not only that, but due to an American news embargo, Japan never even heard that some of its balloons had arrived in the US. But the embargo also meant that the local populace were never warned that they might be at risk. On 5 May 1945, the forests of Mount Gearhart in Oregon bore witness to a tragedy.

It was a mild spring day and Pastor Archie Mitchell, who had just taken up a position in the small town of Bly, drove out for a picnic with his wife Elsie and five children from the church's Sunday school. The war was far away. They were looking forward to going fishing in the river and eating the chocolate cake Elsie had baked. She walked on ahead with the children and had just called back to her husband to say that she had found a balloon when there was a great explosion. The children died instantly, Elsie, who was pregnant, died not long after. Their story has been almost completely forgotten. Years later, shortly after 9/11, President George W. Bush spoke of victims of war who had died on American soil. The victims of Mount Gearhart were never mentioned.

VI
Nature

There's a lot that slime is capable of doing, but it can't do it everywhere. It's an extremely aqueous material and its form and uses vary wildly according to the medium in which it finds itself. In water, it offers creatures added buoyancy, while on land the delicate bodies of gelatinous sea-dwellers like jellyfish collapse instantly. And slimes dry out quickly in air. Thus, over the course of evolution, many gels have migrated inside the bodies of terrestrial organisms like our own. Slimy amphibians are the exception to the rule, but not really: many of them spend their adult lives on land but prefer dark and damp habitats close to bodies of water. Plants use slimes as well, but mostly hidden in the ground, as an interface with soil and its microbes, but also to disperse their seeds – or to catch tasty insects. It is impossible to count or present the innumerable ways in which nature makes use of slime. But a few examples can show the fascinating variety of slimes there is – and why and where they are doing what they are doing.

17

The Magic of the Medium

Life grew and grows. Age by age through gulfs of time at which imagination reels, life has been growing from a mere stirring in the intertidal slime towards freedom, power and consciousness.

H.G. Wells, *A Short History of the World*

Picture the scene: strong men in uniform wielding thick hoses, spraying as far and high as possible. If you're imagining firefighters, you're right on the money. In the 1960s, the New York City Fire Department experimented with adding special agents to their tanks to allow them to shoot the water further under the same pressure. The additive was Polyox and its main purpose was to reduce friction. The natural prototype for this chemical is a viscous mucus found in nature, which is secreted by predatory bony fish like perch and barracudas to reduce frictional resistance by more than half, enabling them to swim particularly fast. It reduces the viscosity of the water at the interface between the fish and its environment. In other words: slippery when wet. The New York trials were deemed a success, but test runs were abandoned all the same. The men in uniform complained that the water spray was too slippery; every puddle was a hazard.

Not all aquatic discoveries translate as easily into terrestrial contexts, but evolution was left with little other choice. After all, life emerged out of the water and we, land-dwelling animals, still bear the legacy of those beginnings today. The first modern blueprints for organisms appeared in the Cambrian Period, over

540 million years ago, spawning predators and species which conquered the open water. Microbes and other lowlier creatures were long established on land when the first marine vertebrates ventured on dry land, as were plants. As self-sufficient photosynthesizers they led the journey inland, and there is a theory that they were helped by fungi which acted as anchors and a support system till roots were invented. Plants were followed by herbivores, with carnivores hot on their heels. These pioneers colonized different habitats and split into species which over a long time adapted to one another and new co-operations. One such example are flowering plants, which require insect pollinators and repay them for their service with nectar. Eventually, these multi-layered interactions gave rise to complex ecosystems on land.

We know of five great mass extinctions in the course of the Earth's history, which wiped out a great number – if not a majority – of the species in existence, reshuffling life's deck of cards. The greatest catastrophe took place during the Permian Period, marking the end of the Palaeozoic era 252 million years ago. Volcanic eruptions, extreme climatic fluctuations and a massive rise in sea levels led to the extinction of nearly 96 per cent of species. Suddenly, many habitats were back to square one. Yet each death of a creature opened up new opportunities for others. Habitats were colonized by competing or newly arrived species, which were able to spread and grow until, perhaps, the next catastrophe put an end to their reign and created the space necessary for other species to establish themselves.

After the Permian Period, reptiles took over. To judge by their grip on the popular imagination, you might think their reign continues. No other group of animals has seized the public's interest as much as the inhabitants of *Jurassic Park*, the dinosaurs on land, as well as the reptiles in the air and sea. In reality, their dominion did not last long. The dinosaurs came, grew and conquered until a meteorite fell to Earth, changing the environment so fundamentally that only very few groups survived. Their loss during the Cretaceous

Period was the mammals' gain. Over millions of years, this once-unremarkable group produced a peculiar, hairless ape, a species that would dominate everything. *Homo sapiens* has a long and winding history, littered with milestones and diversion, and we still carry evidence of it deep inside our bodies today.

No species emerges out of nowhere. We all use blueprints and mechanisms which rarely fit together seamlessly, because they were first used by our ancestors to solve different problems under different circumstances. In evolutionary terms, human beings are four-legged animals which insisted on walking upright, enabling them to explore nearly every habitat on land. In other words, it's our own fault if, sooner or later, we're plagued with back pain and osteoarthritis. We can hardly blame evolution for these deficiencies; after all, it's not all that long ago that it was converting a fish's internal framework into a skeleton, which helped drag the first four-legged creatures on to land and which now keeps us upright on two legs.

But it's not only a matter of mechanics: any creature that conquers a new habitat must overcome numerous obstacles in order to adapt, and the transfer of aquatic organisms to life on land presents a set of great challenges with some concerning slime. Their physiology, behaviour and way of life will have to change because aquatic habitats are shaped only in part by the same forces as terrestrial ones. In short, water is a fundamentally different medium from air.

Let's take a closer look at that medium: water is about 1,000 times denser and fifty times more viscous than air. Anyone wanting to make headway in water must invest the requisite energy and, ideally, use a lubricant. Many fish are muscular, streamlined and slimy, making them good swimmers. They differ from plankton, which generally drift along the path of least resistance. Plankton comprises a wide variety of life forms, microbes such as bacteria, worms, gastropods and gelatinous species like jellyfish, ctenophores and tunicates. What unites them as plankton is their mode of transportation. They float passively through the water or swim with the currents and tides. This saves energy, but it's dangerous because great numbers of predators simply let their helpless prey drift with the flow into their sticky slime traps. One of the few examples of this on land is the spider, which weaves its transparent webs across corridors frequented by insects.

Let's consider the forces at play: all life forms are subject to one constant. Gravity affects us all in the same way and all the time. In the water, however, there is added buoyancy to counter its effect and to keep delicate plankton afloat. Land-dwelling animals must resist gravity, often with supportive structures such as a skeleton. Otherwise, we might break down under our own weight which is what happens to gelatinous sea-dwellers such as jellyfish. Washed up on shore, they go from ethereal elegance to shapeless blob. Fish out of water don't fare well either, even if their bodies are robust enough. Here it is the delicate and slime-covered gills that collapse and dry out in the air, which lets these animals suffocate in a medium which contains much more oxygen than water. And what about currents, tides and waves? Many species like starfish and mussels secure themselves while moving or settling with the help of high-functioning slimy glues – one of the most exciting fields in which scientists can play copycat.

Now, let's take a closer look at ecosystems: since air is drying, land-dwelling animals must regulate and reduce water loss. This limited the first aquatic species to appear on dry land, and amphibians are still bound to damp habitats. Their skin is porous to both gases and water and most species must keep it hydrated with slime. Reptiles went a step further, developing thick scales as protection, allowing them to live in extremely dry regions as well. As mammals, our skin is such an effective barrier that we need not limit ourselves to dark and damp habitats, nor coat ourselves with scales or in slime – at least not on the outside. Some functions, such as digestion or the protection of our less than 'skin-tight' borders in general, still require extensive layers of slime. By necessity of life in dry air we have tucked these away inside our bodies.

What about breathing? Animals depend on oxygen, which cannot be stored in the body and must be constantly replenished from the surrounding medium, either from water or air. Evolution has improvised various mechanisms for exchanging gases, be it an amphibian's skin, an insect's tube-like trachea, a fish's gills or our lungs, which – incidentally – developed in the bodies of fish as well. Some animals combine these features, and changes may also occur during the life cycle: tadpoles breathe through gills before maturing into adult amphibians which take up oxygen through their skin and lungs. It's a dramatic shift connected to the transition from an aquatic habitat to dry land.

In all gas exchange systems, however, one aspect remains the same: oxygen is absorbed by large surfaces and directly into the blood. For this to happen, the oxygen has to be dissolved in liquid. This poses no problem to aquatic species, but what happens on land? Our lungs, and those of many other animals, are coated in slime, with the exception of the alveoli, the air sacs whose thin membrane oxygen has to cross in order to be absorbed into our blood. Here, even a thin layer of mucus would probably hinder the transfer. Instead, another hydrogel barrier coats the alveoli. It's a glycocalyx that carries a layer of surfactant which lowers the resistance for oxygen to enter the body. It also keeps the delicate alveoli from collapsing in on themselves. But why do we have to breathe the oxygen deep into our bodies, with pathogens potentially hitching a ride? Why can't we erect that barrier closer to the surface? According to one estimate, if our lungs were built like external gills, we would lose an impossible-to-replace 500 litres of water per day.

Ex ovo omnia: all animals come from eggs, as the English physician William Harvey claimed in the seventeenth century. Admittedly, he had no proof, and the origins of more lowly creatures were still being traced back to spontaneous generation in slime and mud. During the same period, for example, Izaak Walton's *The Compleat Angler* reported that frogs morphed into slime over the winter, changing back in the spring. It would be a further 150 years before the egg's role in human reproduction was proved by the German zoologist Karl Ernst von Baer.

So how did William Harvey, physician to the king, famed for his discovery of the circulatory system, come to such a perceptive conclusion? He meticulously researched chicks developing inside chickens' eggs, as well as other embryos, finding many similarities. The early development of birds must have been replicated in a similar way in other organisms, Harvey concluded. The principle stayed effectively the same, but not the surrounding medium. How did eggs adapt to the change from water to air? It all comes back to slime.

Cape Cod is a peninsula in the Atlantic, curving like a hook to the south of Boston. On my visit one afternoon in May, the tide is out and mussels lie scattered on the sand next to a dead flatfish, while a red bristle worm with impressive pincers squirms to bury itself in the sand. But I am here for the great moon snail, a predator that lives off the coast, boring into the shells of other molluscs to eat them. I don't expect to see any live specimens but want to find their spectacular egg cases, so-called sand collars. In order to produce them, female moon snails bury themselves in the soft sediment where they mix thousands of tiny eggs with grains of sand, held together by a gelatinous matrix. The slime seems to surround the snail's house like a collar, hardening into a curved shape, sometimes with a scalloped edge like a plumber's fancy plunger. Just the rubber part without the handle, but the image still fits: when I at last find a few sand-collar fragments they are so elastic that I can easily bend them without doing any damage.

Sand collars are unique pieces of reproductive art, and yet moon snails have one thing in common with other marine species like the fish, mussels and worms on the beach this day: their eggs are coated in a gelatinous layer, mainly for protection while letting oxygen through. Can the gels be fitted to purpose? Antimicrobials are a likely addition, as are stronger deterrents. Many sea slugs pack their eggs in gelatinous ribbons that could not be more brightly coloured, from red to pink to yellow. Are they inviting predators to a tasty meal? It's more likely that the eye-catching designs warn of repellent, if not toxic substances. Sometimes – as in the case of certain marine worms – it's the eggs that are colourfully shining out of their clear jelly coating. Corals, on the other hand, release gelatinous bundles of egg cells and sperm. Once they have risen to the surface of the ocean they will break apart and join together with new partners. Mussels simply release their sperm and egg cells without much ado, but covered in a gel-like coating.

Gelatinous layers seem to be the first, if not only, choice for protection of reproductive products. On contact with air, though, they quickly dry out. This is why most amphibians, which spend their adult lives on land, must lay their jelly-coated spawn in freshwater. Some species will choose a watery spot on land but, according to one study, they then invest more time and energy in tending to their brood. Their behaviour can border on the bizarre. Glass frogs use their own urine to keep their eggs moist, while others carry the spawn, and later the live young, around with them. This requires a great effort, sometimes taking weeks, and is intended to ensure the survival of offspring even under particularly difficult conditions.

Reptiles have found another way. They were the first creatures to live their lives where water is scarce. As a consequence, around 340 million years ago, they learnt to cover their eggs in hard shells of calcium carbonate to keep them from drying out. Their cousins, birds, do the same. Just under the shell lies the delicate amniotic membrane, protecting the embryo and its yolk, its package of provisions. But there's another crucial element that distinguishes eggs of this kind: the jelly-like and very viscous albumen, or egg white. 'There is a distinct "nothingness" about albumen,' the ornithologist Tim Birkhead at the University of Sheffield writes. 'It is the most unobtrusive part of a bird's egg . . . albumen seems to be structure-less: a poorly defined glob of mucous-like material.'

The albumen is 90 per cent water, a kind of hydrogel, and as usual not boring at all. 'The truth is that, far from being unimportant, albumen is absolutely remarkable, mysterious stuff,' Birkhead adds. And it turns out that in this case the 'distinct "nothingness"' is of the essence: albumen provides water and proteins for the embryo while keeping it from physical danger. 'But much more crucially, the albumen provides a sophisticated biochemical firewall against the microbes that, given half a chance, would consume the developing embryo,' Birkhead writes. The whole egg is a succession of barriers, from the shell to the densely woven membrane underneath to the albumen as the most important one. It's the high water content – with the nutrients safely tucked away and antimicrobial proteins acting as tripwires – that leaves microbes high and dry. Crossing the albumen to get to the embryo is, in Birkhead's words, like a journey across the Atacama Desert for humans. It's a long and desperate trek.

And what about mammals? We go one step further and keep our young in the womb, where they are fed via the umbilical cord and placenta. Our offspring are also surrounded by a watery buffer, the amniotic fluid. Live birth, as seen in humans, often seems to be peculiar to mammals. But there are also quite a few snakes, such as the common European adder, that hatch their young from eggs inside the mother's body, who then 'gives birth' to them. Other reptiles and even amphibians share this trait – presumably when keeping the offspring safe and secure at all times might be the better option. In terms of evolutionary change, to go from egg-laying to live birth is rather complicated, with many single steps involved. But at least one species in transition seems spoilt for choice: the Australian skink lizard *Saiphos equalis* was known to give birth to live young, but one female was observed laying eggs one week and giving birth to live young a few weeks later. If it is not an evolutionary transition, these animals might be able to choose between two kinds of reproduction depending on environmental conditions.

18

Life Under Water

'. . . so that at 32 feet beneath the surface of the sea you would undergo a pressure of 97,500 lbs; at 320 feet, ten times that pressure; at 3200 feet, a hundred times that pressure; lastly, at 32,000 feet, a thousand times that pressure would be 97,500,000 lbs – that is to say, that you would be flattened as if you had been drawn from the plates of a hydraulic machine!'
'The devil!' exclaimed Ned.

Jules Verne, *Twenty Thousand Leagues Under the Sea*

Nineteenth-century Victorian society was seized by a succession of unlikely crazes, one of which was a passion for ferns. It was the age of discovery and everyone wanted to decorate their homes with wondrous living pieces. Once the botanist Nathaniel Bagshaw Ward had developed a sealed glass container for living plants, everyone from the humble labourer to the aristocrat could indulge in 'pteridomania' – tending to, growing and studying ferns. The obsession went so far that some species were pushed to the edge of extinction, while elsewhere new hybrids emerged out of the contrived proximity of different species.

These tightly sealed glass cases also opened the public's eyes to the previously unknown wonders of the oceans. The first scientific expeditions had brought up mysterious creatures from the deep such as unknown corals, crabs and sponges. They could hardly be studied in their natural habitats, but their new homes behind glass made observation possible, as Anna Thynne, the first person to keep sea creatures in an aquarium in London, successfully proved. She kept marine animals like stony corals alive for years, allowing them to thrive and even multiply. It was no easy undertaking, particularly for her staff. At least once a day, a maid was required to spend a quarter of an hour aerating seawater, which had been transported to London from the coast, by standing next to an open window and pouring it from one container into another.

Thynne published her findings with the help of Philip Henry Gosse, who would go on to write a popular manual on the new craze, introducing the hobby to the masses. People keen to keep abreast of the latest fashions could now enjoy a little piece of the underwater world in their own living rooms. The big cities followed suit and huge aquaria were built as public attractions, from London to Berlin to New York. But exotic wildlife proved tricky to domesticate. Back then, it was simply impossible to keep conditions in the aquaria constant, oxygenation remained an issue and sensitive creatures died off by the dozen.

Soon enough, the public lost its enthusiasm for aquatic exhibitions, which often saw unhappy sea life wasting away, bobbing up and down in murky water. Yet an interest in the oceans survived. It was the age of marine discovery, and HMS *Challenger* was leading the charge. Marine biologist Antje Boetius, of the Max Planck Institute for Marine Microbiology in Bremen and the Alfred Wegener Institute in Bremerhaven, and her father, the writer Henning Boetius, take stock of the *Challenger* expedition in their book *The Dark Paradise*. The voyage, lasting four years and covering nearly 70,000 miles, involved 734 deep-sea mapping explorations and 255 deep-sea temperature recordings, and the trawler nets were deployed 240 times, creating a first, if fuzzy, picture of the oceans and their currents. This included thousands of species of sea creatures hitherto unknown to science.

The findings finally put to bed the Abyssus or Azoic theory, which speculated that the deep ocean was a dead zone devoid of life below a depth of 550 metres. There were also, according to the Boetiuses, numerous 'fantastic strokes of luck', such as a reading on 23 March 1875, in which the lead line spooled down seemingly endlessly close to the Pacific island of Guam, only reaching the sea floor at a depth of 8,100 metres: 'like they had discovered the gates of hell'. The spot was named Swire Deep after Herbert Swire, the sub-lieutenant on board. It is part of the Mariana Trench, an oceanic trench which is also home to the Challenger Deep and, at a depth of almost 11,000 metres, is the deepest place on the planet.

Deep underwater, eternal darkness reigns. Temperatures are close to freezing and the pressure is 1,000 times higher than at sea level. The region below 6,000 metres is known as the Hadal zone and is reminiscent of the kingdom of Hades, god of the Greek underworld. Animals should hardly be able to survive here, at least that was what we thought until scientists took their machines into the hellish depths for the first time. A few years ago, they encountered to their great surprise a creature that, like so many others, uses jelly-like materials to adapt to its marine habitat.

The hadal snailfish lives in the Hadal zone in the North-west Pacific, where it can be found swimming rather too busily to do its name justice. It is a member of the snailfish family *Liparidae*, of which several hundred species of various different colours are already known to science, many of them inhabiting the world's deep-sea trenches. The species *Pseudoliparis swirei* – also named after Sub-lieutenant Swire – lives more than eight kilometres below sea level and holds the record for the world's deepest-dwelling fish. It's an astonishing feat for this lively, pink-bodied little creature; the water pressure at the point where the fish was found is equivalent to the weight of an elephant on a fingertip. How do these animals compensate for the significant pressure they are subject to in this habitat?

With its bulbous little body flowing into a flat, undulating tail, the scaleless snailfish looks like an oversized tadpole. It is faintly translucent, due to the gelatinous threads which run through its tissue. This jelly-like matrix helps it to withstand high pressure, improves buoyancy and probably makes it more streamlined. Many deep-sea fish produce gelatinous material of this kind, highly hydrated matter which requires little energy to construct while offering a quicker way to build up body mass than muscle would. This technique only works under pressure, however: if the snailfish is brought up out of the ocean depths, its tissue melts. The similarly gelatinous blobfish (*Psychrolutes marcidus*) was declared the world's ugliest animal in 2013, even though its grumpy expression in a lumpen face was simply due to its collapsed tissue.

Here's a curious side note: terrestrial organisms offer an unlikely champion of jelly-like structures in their tissues – plants. These gelatinous fibres or G-layers might have evolved with early land plants and are still widely used. The best-known example is that of trees which use gelatinous fibres in their aptly named tension wood to make sure their stems grow and keep upright while giving the branches a different orientation. Gelatinous fibres contain a sugary matrix and exhibit gel-like behaviour like shrinking and swelling. This could in itself be a desirable function, since it conveys some flexibility to otherwise rather rigid plant structures like stems, branches, thorns and tendrils. Or even to whole plants: in some cases, these fibres pull entire shoots underground to survive fires or freezing.

But let's return to the sea, where gel-like bodies are not confined to the deep ocean. Jellyfish, ctenophores, tunicates and many other animals – including planktonic larvae of myriad species – are composed largely of gelatinous matter. The bodies of jellyfish and comb jellies are made up of gel-like mesoglea, elastic fibres as well as muscle bundles and nerve fibres embedded in a highly hydrated matrix.

This is what makes the common jellyfish or moon jelly, *Aurelia aurita*, one of the ocean's most effective swimmers, as the biophysicist John Dabiri of the California Institute of Technology was able to show. The animal's bell pulsates, which rolls water off the top of it, creating a pull at its apex which the jellyfish uses to propel itself forward in a kind of suction movement, requiring no additional energy. A recent publication proved that the animals use another physical force to their advantage: when a plane lifts off or an animal swims close to a solid boundary, the so-called 'ground effect' gives them extra push. Jellyfish swim in open water without any natural surface in sight. But *Aurelia aurita*'s movements create a vortex in the water that acts like a 'virtual wall'– making the master swimmer even better.

It's an astonishing degree of efficiency for an animal that is made up of cheap biological material. The common jellyfish is little more than water, which offers a crucial advantage in the open ocean, though. These blue deserts, with no stony areas of shore, forests of kelp or other forms of hiding place, leave prey animals vulnerable if they don't adapt to their environment by becoming invisible. Members of various groups have shifted to using gooey bodies because the material more or less reflects and bends light like its surrounding medium. It looks and behaves like water in the open ocean; it is, in other words, transparent. But not all body parts are capable of that. Eyes need to reflect light, and the digestive tract will be visible, at least when it's filled. That is why one type of camouflage is not enough.

The hyperiid amphipod *Cystisoma*, a marine crustacean, for instance, can grow as long as a hand and is almost invisible. It helps that the animal has huge but only lightly tinted eyes because the dark, pigmented cells are spread over a large area. The trick works, as the biologist Karen Osborn at the Smithsonian Institution explains: 'Mostly you see them because you don't see them. When you pull up a net full of plankton, you see an empty spot – why is nothing there? You reach in and pull out a *Cystisoma*. It's a firm cellophane bag, essentially.'

The glass squid goes even further. Its body is see-through but there are, again, those potentially treacherous eyes and the dark gut.

Most predators approach from the depth and scan the water above them against the sky to find prey, but they will be hard-pressed to make out the squid. This time the animal seemingly fights fire with fire by illuminating its own eyes. However, this is no highlighting, but counter-illumination to hide any hard contrasts. That leaves the digestive gland as a problem to solve. This organ works a bit like our liver, is cigar-shaped and dark – and it can spin. As the squid moves around, the gland remains constantly upright, like a kind of biological compass needle. Hunters peeking up from the ocean's depths, trying to find their prey, will have to spot the needle-like tip of the organ.

A couple of terrestrial species also make attempts to disappear into thin air, including the glass frog, whose camouflage is better described as translucent than transparent, according to a recent publication. This is not see-through invisibility but a softening of the edges, the blurring of a silhouette in order to meld visually into its surroundings. And there is a reason why gelatinous bodies give terrestrial animals away: highly hydrated jelly imitates water to perfection because it is not much more itself. But gelatinous bodies fail to imitate the less dense air, which bends and reflects light in a different way – which is always a giveaway. Even though the dream of invisibility is as old as humanity, living creatures will probably have to rely on optical tricks instead of actual see-through bodies since these would have to behave like air.

H.G. Wells must have pondered that problem a lot, preferring as he did to underpin his novels with solid science. In *The Invisible Man* he sets himself the task of describing the transparent body of the scientist Jack Griffin – the result of a failed experiment undertaken in a fit of hubris – in a manner both plausible and consistent, right up to the piece of cheese the scientist eats, which consequently makes its 'ghostly' way through his invisible digestive tract:

> Is there such a thing as an invisible animal? . . . In the sea, yes. Thousands – millions. All the larvae, all the little nauplii and

> tornarias, all the microscopic things, the jelly-fish. In the sea there are more things invisible than visible! I never thought of that before. And in the ponds too! All those little pond-life things – specks of colourless translucent jelly! But in air? No!

Wells did a good job coming up with a sciencey explanation for his hero's transformation that is completely unrealistic at the same time. True transparency will, for the time being, be the privilege of gelatinous animals in the sea, all but looking like water themselves. But see-through bodies aren't the only tricks they have come up with to hide from predators. Slime can help in other ways too.

A slime-screen is one possibility. Some marine snails like the sea hare *Aplysia* emit purple clouds to ward off predators, with toxic ink as the main ingredient. The dark cloud is kept from diffusing right away by a good dose of mixed-in slime. Again, some squids go one better. If in danger, they add enough slime to their ink to create a pseudomorph. These are squid-shaped and squid-sized doppelgängers with only one job: to stay stable long enough to distract the predator. One species is even capable of creating a whole army, shooting out several pseudomorphs in a row, before blending in discreetly among its slimy comrades or slinking away.

But employing slime as a distraction doesn't always have to be a matter of life and death. All the parrot fish really wants is a good night's sleep out on the reef. Is that too much to ask? Without proper equipment it would be, but the brightly coloured animal simply secretes a slimy balloon to hide in. The head-to-tailfin sleeping bag is see-through but is thought to keep telltale molecular scents from escaping, which makes the fish all but invisible to parasitic *Gnathiidae*, the sea-dwelling equivalent of ticks.

Should these, or some other pest, latch on nonetheless, the unfortunate victim need only swim by a cleaner station in the coral reef. Big fish, turtles and even octopus can stop by to have dead skin and external parasites picked off by sharp-toothed cleaner fish. Mutual trust, or at least a truce of some sort, is essential because these little helpers work inside their clients' open mouths, in between their sharp teeth. Yet it seems that the predators fall into a kind of trance which relaxes their biting reflex. This suits the cleaner fish too, because they're able to snaffle small bites of nutritious slime as a treat off the skin of their daydreaming customers. It suits the bluestriped fangblenny even more, a mimic of a cleaner wrasse that only wants to get close enough to rip a mouthful of flesh out of an unsuspecting client, whose response to the attack will still be muted due to the parasite's opiod-based venom.

Grabbing a bite of mucus or flesh is always a challenge, especially if your prey are stinging corals with a razor-sharp skeleton. The tubelip wrasse (*Labropsis australis*) has come up with an ingenious solution by giving a lubricated kiss of death. Its fleshy lips are arranged in fine folds, like the gills on a mushroom, and they are beset with goblet cells that make the mouth ooze with slime. That way the animal can suck the mucus and flesh of corals without feeling their stings or cut up its own delicate flesh. Another example where a fish's soft anatomy is highly adapted to help handle prickly food concerns a type of wrasse that produces copious amounts of mucus in its mouth. Its diet consists mainly of gelatinous food – either organic waste or zooplankton – and the mucus might help to retain the slippery fare and neutralize any stinging cells.

But not all nutritious mucus must be fought for. Discus fish dispense their own slime willingly. Well, for a time at least. Both male and female parents allow their young to spend a month eating the rich gel from their skin, which is saturated with immune factors. However, as the weeks pass, the provisions cause conflict: the young latch on with greater frequency, with parents taking alternating shifts until they eventually go on strike. It's a special way of caring for a brood; scientists consider the biparental mucus-feeding more similar to the habits of mammals and birds than of other fish. And it's not the only example of cannibalistic offspring: caecilians are land-dwelling amphibians whose females allow their young repeatedly to gobble up the thick outer layer of their own skin.

But back to well-armed prey – and shelter: pearlfish hide in an unexpected place, as John Steinbeck observes in *The Log from the Sea of Cortez*:

> In one of the sea cucumbers we found a small commensal fish, which lived well inside the anus. It moved in and out with great ease and speed, resting invariably head inward. In the pan we ejected this fish by a light pressure on the body of the cucumber, but it quickly returned and entered the anus again. The pale, colorless appearance of this fish seemed to indicate that it habitually lived there.

And they need their plentiful skin-slime as a lubricant when they slip into the sea cucumber's rear end, which can't be squeezed shut since these creatures breathe through their anuses. To add injury to insult, *Encheliophis* pearlfish not only use their hosts as refuges but also eat the sea cucumbers' internal tissues. Yet the inside of a sea cucumber is not completely defenceless against attacks of all kinds. It can expel its thread-like and rather sticky intestines, which also secrete powerful toxins. This doesn't make for a cosy shelter, but pearlfish somehow prevail by secreting an extra-thick slime coating for protection.

The pearlfish's slimy sheath might be a unique feature in response to its special accommodation, but external mucus layers also help other fish to lubricate their way through the water and narrow openings. And these barriers possess many more important functions as the interface between the animal and its environment. We know that the mucus can contain antimicrobial and immune-related molecules to prevent infections while housing the microbiota. Fish mucus – which can be similar to our mucin-based hydrogels – has a social function as well. It helps with the communication between members of the same species to synchronize their spawning or co-ordinate shoaling, for example.

Communication is a double-edged sword, though, since it can

lure unwanted suitors as well. The parasitic flatworm *Entobdella soleae* only attaches itself to the skin of the common sole, which their larvae must seek out and infest immediately after hatching. The nocturnal sole spends its days half buried in sediment, which makes it easier to target. This is probably why most attacks happen in the morning, but the larvae keep their schedule flexible. If they so much as catch a whiff of the slime the sole has left nearby or even on top of their eggs, they will hatch immediately.

Scientists have been trying to copy that feat of honing in on mucous markers. They often have a hard time detecting all species, especially the rare or hidden ones that live in aquatic ecosystems. But since sloughed-off mucus can contain cells of the organism it came from, all the scientists have to do now is screen water samples for genetic traces, the so-called environmental DNA. A similar method can be useful for checking the health of giant organisms. Scientists used to be reliant on skin and tissue samples to assess a whale's health, but these were hard to get; now they use drones to catch the mucus that gets expelled whenever the animal breathes through its blowhole. It contains cells of the whale itself but also samples of the microbiota, and possibly pathogens.

Dangerous stowaways are a problem on their outer barriers as well. Many whales are routinely and visibly infested with parasites and other pests, which is a consequence of their unique evolutionary history. Unlike fish that never left the water, whales were adapted to life on land without an outer layer of mucus before they returned to the sea which makes it easier for parasites to attach. Pilot whales, however, have developed a very smooth skin which is self-cleaning. The spaces between their cells produce a kind of slime containing enzymes which fills in uneven spots and makes it harder for pests to gain a foothold.

But it's an eternal arms race, and some parasites may in turn adapt to the new barrier and use it to find their host. Not all slime-loving larvae are a menace, however. The microscopic offspring of worms, mussels, corals, crustaceans, sponges and other marine animals float through the sea as plankton, looking for a good habitat. Since they settle down only once to metamorphose into their sessile adult forms, it has to be the perfect spot. Numerous environmental factors have a part to play in this process, which is critical for the survival of whole populations of marine invertebrates.

When the larvae choose their future homes one aspect stands out, which some scientists see as a universal mechanism. Larval settlement and metamorphoses could be induced – and possibly inhibited as well – by microbial slimes. These complex biofilms are ubiquitous and will quickly grow on any surface in seawater, often with different species of bacteria, unicellular algae and other microbes. It is difficult to unravel which specific signal sends which kind of message to induce or repel different invertebrate larvae, and we don't know the details in most cases yet, but the connection itself is established. Larvae of the tubeworm *Hydroides elegans*, for example, will refuse to latch on if a biofilm is not in place, and even seem to prefer specific bacterial species.

If certain biofilms offer marine larvae 'love at first taste', as some scientists have called it, then sharks get all the feels from slime. Just like rays, these predators hunt with the help of sensory organs in the skin, known as ampullae of Lorenzini. Filled with jelly, these pores and channels pick up the tiniest changes in pressure. If an organism moves even slightly and at a great distance, the shark can locate it via its slimy antennae. If the search leads to a hagfish, though, the shark will end up with only a slimy gag for its trouble. Disappointment is also served up to the unlucky ray that risks a bite of the starfish *Pteraster tesselatus*: under attack, a hollow layer under its skin floods with enough repellent slime to spill over.

Another slime-emitting sea creature is the worm snail (*Vermetidae*). After they settle as larvae, the adult animals spend their whole lives in one spot in chalk tubes that look like either tightly wound or unravelled snail shells. That lifestyle poses two problems: how to feed? And how to reproduce? Slime is the answer to both questions. Like spiders in their webs, the worm snails let sticky lines float in the currents as traps for prey. From the opening of their tubes, they shoot slime nets into the open water that may even overlap like a web in colonies of the animals. These slimy shrouds can destroy coral tissues, which suggests that they may well contain toxic chemicals. When the time comes for reproduction, the males simply release their sperm bundles into the open water where they become trapped in the females' nets, sticking to their slimy fishing lines before being reeled in.

In the dark and rather empty depth of the sea, however, females stuck in one place couldn't risk having their sperm traps come in empty again and again. The worm *Osedax mucofloris* has had to find another way to secure the next generation. This bizarre animal lives on the sea floor absorbing the last nutrients and fats from bones, preferring the skeletons of whales that have sunk down after their deaths in a journey that can last for weeks. These whale-falls induce a kind of spring in the deep sea, where hundreds of species rely on the bounty from above, even if they're not as specialized as *Osedax* is. The worms anchor themselves to the bone tissue using spurs, much like the roots of plants and covered in a mucus that dissolves the tissue or protects the animal amid the crumbling bones. But the whole animal is surrounded by a gelatinous tube which houses a harem of more than 100 dwarfish males.

19

Between Two Worlds

Frog is a soft bag of green.

Anne Sexton, 'The Frog Prince'

An early memory takes me back to our garden, where I'm lying on my stomach on the grass next to my father, between the flower beds and the garage, eye to golden eye with a frog, shockingly green in the fresh spring grass. Its toes end in sticky little pads, its expression is melancholic yet bold, probably due to the black strip which extends from its eyes to its back legs, like Zorro's mask. Did I touch it, or is my memory of its skin delicate as tissue paper beneath my fingers deceiving me? It's a tree frog, a somewhat plain name for this beauty. Yet in German texts it also appears as 'tree-sticker' or 'weather frog' or, better yet, as 'green frock', a tailor-made name for this jewel of a creature. It might even be more fitting not to settle on a single label but let all the names stand, to catch the frog's ambiguous nature.

After all, it is a member of the amphibian family that carries its ambivalence in its name, which comes from the Greek *amphi*, 'both', and *bios*, 'life'. These animals lead double lives, beginning usually as algae-munching, water-dwelling tadpoles before undergoing an extraordinary metamorphosis which sees them transform into land-dwelling carnivores. They retain their aquatic heritage: adult amphibians have an overwhelming preference for cool, damp habitats, as we will see later. In this chapter, we'll take a closer look at a variety of creatures that live between two worlds, using slime to master life in different spheres.

First up, the tidal zone. Animals living here have to cope with extreme fluctuations in their environment on a yearly, daily or even hourly basis, as it either dries out or floods. As a consequence, its inhabitants also have to cope with the extremes of temperature and salinity. As we have seen, limpets and other molluscs retreat during low tide, hunkering down in the rock while sealing themselves off with slime to prevent water loss. Unicellular diatoms, on the other hand, journey even further. These ubiquitous micro-algae construct an ornate cell wall from silica; it's their very own glasshouse. But that striking appearance comes with outsized importance: diatoms produce at least 20 per cent of Earth's oxygen through photosynthesis. To gather sunlight, marine species spend their days near the ocean's surface.

As evening approaches, many of them sink down and let the waves carry them to the coastline, where they use sticky slime to attach themselves to grains of sand. Like sea-weary sailors, they spend their nights on firm ground. In the morning, they shed their slimy anchors and allow themselves to be washed back out to sea again. They will never land in the same spot again, but they do leave their mark. They are so numerous that their sticky goo – together with other microbial glues – helps keep whole stretches of coastline in shape, as well as influencing the movement of sand and mud in other environments like the floors of rivers and the sea. These exopolymers stabilize the beds by sticking the grainy particles together and increasing resistance to erosion.

Low temperatures can be a kind of interface as well, where species have to adapt to an extreme and highly changeable environment. Diatoms of the species *Melosira arctica* live on sea ice and their slime acts as an anti-freezing agent, protecting the single-cell algae at temperatures even as low as -10°C. Secreted exopolymers change the habitat here too, melting channels and pores in floating ice so the algae can live in a cosy microclimate along with other species who find a home here. The Antarctic limpet *Patinigera polaris* withstands similarly low temperatures; it can even be stuck in ice and survive. It does this by coating itself in slime which probably prevents dangerous ice crystals from forming inside its body. Without this gel coating limpets exhibit higher mortality rates, and it remains unclear as to whether the slime's naturally increased viscosity lowers their freezing point or whether special properties play a role. But external slime layers seem to offer protection: the Antarctic ploughfish (*Gymnodraco acuticeps*) is a scaleless fish that secretes glycoproteins with anti-freeze properties which are not found in the blood, as they are in other polar fish, but in the slime on the outside of the body.

And slime even protects future generations. The flightless midge *Belgica antarctica* is Antarctica's only true insect and a master of survival in an extreme environment. As mild days alternate with bitterly cold ones, the black fly, its worm-like larvae and eggs are either freezing or thawing, but only one of them wears a coat. Scientists have found evidence that the females pack their eggs in a thick layer of anti-freeze gel. The mechanism is unclear, but the gel acts as a buffer against extreme temperature fluctuations and also protects the eggs from Antarctica's extreme dryness.

Cold poses no problem at all in the homes of other habitat-hoppers, but close contact with rough surfaces is a challenge. Amphibian fish like the mudskipper live in the mangroves of the tropical tidal zones, where they alternate between water and land. This kind of fish can survive out of water for several days, moving around with the help of their powerful pectoral fins and absorbing oxygen through mucous membranes in their mouths and throats. A layer of gel on their skin protects them from abrasions. For many scientists, this modern-day species bears an echo of the time when vertebrates first arrived on land. These pioneering creatures are descended from bony fish, which had already devised how to breathe through their lungs while still in the water. The amphibian fish of today are not evolutionary relics, but they may offer some indication as to how the first amphibians arrived on land and for a time managed to establish themselves at the top of the food chain without competition.

No doubt there will be many more fish, flies or even freezing limpets that manage to survive where different spheres meet. But as a group, it is surely the amphibians who best embody a life between two worlds. Not all of them try this balancing act at the climatic extremities, but many species in the three amphibian groups – with frogs and toads as one, salamanders and newts as the second and the mysterious caecilians as the third – get creative in order to thrive both in water and on land, usually with slime. Amphibians use hydrogels to catch prey, repel predators, safeguard their eggs – and even provide an erotic boost for the males during mating.

A crucial amphibian slime is the outer mucus layer because it covers a skin that is unique among vertebrates for being both robust and somewhat porous. Amphibian skin helps with camouflage, protection from UV light, defence, respiration, water regulation, thermoregulation and communication with members of the same species. The mucus layers helps with this too, but its most essential function is for water conservation and gas exchange. As an interface with a world full of pathogens, the slime both houses the microbiota while also being fortified with antimicrobial and immune-related secretions and molecules.

And some amphibians can ramp up their mucus production when attacked. The 'snot otter' even oozes with stinking slime when irritated. The giant hellbender salamander is a descendant of *Andrias*, the model for Karel Čapek's fictional newts that went to war with humanity out of necessity, and this modern-day salamander only wants to be left in peace as well. Although its smelly mucus might keep predators away, it may have inadvertently attracted human interest, for the repulsive gel has potential uses in medicine for its ability to help seal wounds and encourage fast healing, even though this has only been demonstrated in animal models so far.

Amphibians' slime layers fulfil a double function, forming a powerful barrier to protect their vulnerable skin, while allowing water and gases to pass through. This system works so well in frogs that they can spend entire winters underwater, which is probably how they got their reputation as seasonal shape-shifters that melt to slime late in the year only to be reconstituted in spring. '[T]he land frogs are some of them observed by him, to breed by laying eggs, and others to breed of the slime and dust of the earth, and that in winter they turn to slime again, and that the next summer that very slime returns to be a living creature; this is the opinion of Pliny,' wrote Izaak Walton in 1653 in *The Compleat Angler, Or Contemplative Man's Recreation*.

Toads' thicker skins do not exchange gases underwater as efficiently and females can drown during mating when too many males try to cling on to them at once. Their less slimy skin has another disadvantage: they are rarely too slippery for predators. It is thought that they may compensate for this by producing potent toxins from glands on their skin. This behaviour is also exhibited by the rough-skinned newt (*Taricha granulosa*), which lives on the west coast of America and is actually a salamander. It secretes tetrodotoxin (TTX), which also poses a risk to humans.

Their toxic skin makes them reckless. One gloomy, rainy day I'm in a forest in Oregon at twilight when one of the newts almost crawls under my shoe, its dark body practically flush with the ground. But instead of running away, it tries to frighten me off, arching its back to present its bright-orange belly. This is the unken reflex, an unmistakable warning. It might succeed at keeping me at bay, but the same move does little to discourage garter snakes, their primary predator. The snakes don't die from TXX but slow down or become paralysed, and it seems that some of them don't suffer even these effects.

They have become resistant to the deadly neurotoxin, whose dosage the newts have cranked up in return. But it's not the amphibian who has to work extra shifts here: TTX is too complex a compound to be produced in the newt's body and is instead a gift from bacteria living on the newt's skin, members of its microbiota. The evolutionary struggle is still ongoing, but it seems that in some places the snakes have won – and the least resistant reptiles can withstand even the most toxic newts.

The common death adder of northern Australia is also capable of thwarting several defensive measures on the part of its amphibian prey: all it takes is a little patience. When attacked, the marbled frog *Limnodynastes convexiusculus* secretes an extremely sticky slime, while Dahl's aquatic frog (*Litoria dahlii*) emits a lethal toxin. Yet the death adder simply strikes and waits; the sticky slime will harden and the toxin will decompose, albeit at a different pace, and the snake seems to know this. It waits just ten minutes to eat the marbled frog, but it's a solid forty minutes before it tucks into the aquatic frog. The predator's response to the amphibians' defence systems has undergone an evolutionary upgrade, such that it might now be able to keep the upper hand.

For all their defensive tricks, amphibians are the most threatened vertebrate class. Their most deadly foe is not a fellow creature, but the frequently fatal skin disease *chytridiomycosis*. This disease, along with other factors like habitat loss and climate change, is leading to the amphibians' global decline. It's caused by two fungal pathogens with different targets: *Batrachochytrium dendrobatidis* (Bd) infects frogs while *Batrachochytrium salamandrivorans* (Bsal) kills salamanders. As one recent publication states: 'The severity and extent of the impact of the infection caused by these pathogens across modern Amphibia are unprecedented in the history of vertebrate infectious diseases.' The animals are water-dependent, which is one factor that can favour the growth of any number of fungal pathogens – even pandemic ones.

Yet the amphibians' first line of defence against pathogens is the skin and its outer mucus layer. And they have powerful defences that often seem to keep the invasive fungi at bay. The microbiota are thought to play an important role as well, either by competing with the invaders for nutrients, by secreting powerful antimicrobials or by influencing the host's immune response. Some of the microbes and their products look like key players here and probiotic therapies are being tried at the moment: if successful, threatened amphibians' defences could be artificially fortified through the application of specific bacteria and their compounds. Other approaches are being tested as well, and the amphibians themselves might secure their own survival. It seems that some species recover from the brink of extinction, possibly by adapting their immune responses to the fungal threat. But even if that will turn out to be true, much will have been lost on the way.

Australian gastric-brooding frogs, for instance, are a curiosity even among amphibians and were some of the first to disappear, along with their unique lifestyle, which saw the females swallow the eggs and incubate them in their stomachs, eventually vomiting up the live young with gusto. Inside their temporary womb, the offspring produced slime which reduced the production of stomach acid. Research in this area had promised new ideas for potential therapeutic antacids in humans, but the animals disappeared in the mid-1980s before this was possible. Australian scientists started the Lazarus Project in the hope of bringing these creatures back to life. The DNA of gastric-brooding frogs has been transferred into the ova of a related species. To date, however, no tadpoles have hatched successfully. Could it be that they miss the telltale scent of home?

During mating season, most frogs and toads are drawn to the site of their birth and rarely travel to other bodies of water. Perhaps it has something to do with an enticing scent in the air which they pick up, smelling, tasting and sensing it through their sensitive skin. By the time they arrive, the males will be clinging tightly to the females in a phenomenon known as amplexus, clasping their bodies in an iron grip. If their arms are too short, as in the case of the African common rain frog (*Breviceps adspersus*), the males simply stick themselves securely to the females' backs using slime. Animals looking to mate aren't particularly choosy when it comes to romantic clinches, however; even fish have been known to be mounted, and there are images circulating online of an Australian python boasting no fewer than ten male toads on its back.

Once couples have paired up, fertilization takes place in the water and must be synchronized outside the body. Female toads produce two long strands of eggs and the males – still riding piggyback – ejaculate as soon as they feel the gelatinous egg cases touch their toes; evidently that's what turns them on. To be fair, though, as far as unusual erotic preferences go, slime seems to have some universal appeal and capitalizes on sensual kicks of the sticky, gooey and viscous kind. In but one example there are slimy women whose whole bodies are not only see-through but also 'slowly pulled away from each other, strings of their essence popping away from each other', as described in *My Sexy Slime Girlfriend*, a novel by Aya Ocean: 'It means I want to stick to you, and I want you to stick to me. Does that make sense?'

For amphibians, of course, it's less about romance than synchronized reproduction when a tightly entangled pair of toads paddles frantically in order to wrap the strands of spawn like strings of dark pearls around the stems of aquatic plants. If the female takes a break, the male holds off until he feels her irresistible jelly again. For frogs it's a less leisurely process, as females lay their spawn all at once. Its gelatinous coating swells swiftly in contact with water, becoming firm – in some cases accidentally producing star jelly. If all goes to plan, the males ejaculate swiftly and in one go while wildly paddling to mix the sperm with the eggs before the gel hardens.

Once the tadpoles hatch from their jelly, as not much more than little dark blobs with a tail, they begin life as aquatic herbivores. But eventually they swap their tails for legs to hop on land as adult frogs or toads and become meat-eaters. There's no hurry, though, as the amphibians often don't even have to out-jump their prey. Instead, their tongues shoot out like a coiled spring, stretching a full body's length. It's the perfect glue trap for somewhat lethargic predators.

The tongue itself is as super-soft as brain tissue and during impact its tip will nestle up against the prey. That way it can bring its own saliva in even closer contact with the hapless victim. It's a viscoelastic slime with a rather liquid consistency that will allow it to cover even wet worms, hairy mice or armoured insects perfectly. As the tongue springs back, the viscoelasticity changes and the by now extremely sticky saliva has enough grip to hold the prey in place for its bungee jump of doom. Having the snack firmly stuck in the mouth would lead to some sort of stalemate, though, if frogs hadn't come up with an ingenious method to free the trapped creatures from their extremely sticky tongue.

When swallowing, frogs close their eyes like the most self-indulgent connoisseurs. Some species retract their eyeballs ever so slightly, using the backs to shear the prey from their tongues into their throats. Viscoelastic hydrogels work as saliva, as long as the frog keeps an eye on them . . . or maybe two. This technology is sophisticated, and all the amphibian sticky traps should guarantee frogs, toads, salamanders and reptiles like chameleons the upper hand over insects. Why, then, do the large larvae of the ground beetle *Epomis dejeani* parade provocatively in front of amphibians? Because it's a predator who has learnt to evade ballistic tongues and latch on to frogs or toads – and suck them dry.

Amphibians may be slippery and toxic, but they are highly desirable prey. To counter the threat of multiple predators they have numerous offspring, which might seem an unnecessary indulgence at first glance. It's a bitterly cold April day in the Scottish mountains, just tentatively green, and yet thousands of jelly-like eyeballs with tiny pupils are staring at me from every stream, rivulet and puddle. I can pick up the masses of spawn in my cupped hands and feel the weight of it, heavy in my palms. The parent newts are dainty little things, tricky to spot amid all their eggs which seem to replace the water in some spots.

It should be much too cold for tadpoles to develop, but it is thought that the thick jelly buffers the eggs against the icy temperatures or might even keep the embryonic black dots warm, magnifying the sunlight inside. And there's no time to lose; a fast development can improve the embryos' chances of survival, which is not at all assured. What looks like an overabundance of spawn will dwindle quickly. Only 5 per cent of amphibian young will eventually leave the water and only a fraction of those will return when they reach sexual maturity, to produce the next generation. This has been demonstrated in studies on some species of amphibians, and the results can very likely be applied to others.

Devoted parenthood seems the only alternative to that kind of terrible waste and one amphibian group takes it to an extreme. Secretive caecilians shy away from quantity when it comes to their offspring, but give their all to the few young they have. These tropical legless creatures look a bit like wormy snakes with sharply pointed teeth. They breathe through the skin and mucous membranes in their mouths. Some species lay gelatinous eggs in the damp earth, but most caecilians give birth to live young. But while still in the womb, *Boulengerula taitana* young will scrape at the lining of the uterus to stimulate the secretion of what is known as uterine milk. They remain with their mother after birth, repeatedly skinning her over the course of several weeks. These baby cannibals bite the female's thick outer layer of skin and spin along her long axis, tearing off one fragment of skin after another.

This generosity on the females' side is not extended to predators, though. Like all amphibians, caecilians produce slime containing a tailor-made cocktail of active agents. The Brazilian species *Siphonops annulatus*'s ground-based way of life requires an asymmetrical distribution of glands. The slime produced on its head is primarily a lubricant, allowing the animal to slip into the earth much like the root tips of plants. Toxins matter more at the tail, which seals the caecilians' ground nests like a cork in a bottle and scares off poisonous coral snakes the moment they take a first bite. But these functions are not as strictly separated as was first thought. *Siphonops annulatus* and other caecilian species possess specific glands at the base of their teeth that give them a venomous bite – which could make them the oldest venomous creatures on Earth.

20

On Dry Land

> *Turdus ipse sibi cacat malum.*
> *The thrush himself excretes his own trouble.*
>
> Latin proverb

Baldr, the god of beauty, light and happiness in Norse mythology, met a rather unhappy fate. He was brought down by mistletoe. The goddess Frigg extracted a promise from all the living things of the Earth not to harm her son Baldr. However, cunning Loki discovered that Frigg had overlooked one plant, perhaps because it wasn't anchored in the ground. Parasitic mistletoe sinks its roots into a tree's tissue, robbing it of water and nutrients. Common mistletoe is native to Europe and is hemiparasitic: it photosynthesizes through its evergreen leaves and produces its own sugars.

The mistletoe's white berries are sometimes referred to as Frigg's tears, those cried for the loss of her son, as Jonathan Briggs writes in *A Little Book About Mistletoe.* The goddess didn't bear a grudge against mistletoe, however; instead she declared it a symbol of love, which is why, to this day, we kiss under it at Christmas time. Scientists today are still unsure how this evergreen parasite continues to thrive in winter, even when its host tree has lost its leaves, but since the time of the Celts mistletoe has been believed to have magical properties. In *The Adventures of Asterix*, the popular French series of comic books, a Gaulish village fights the Roman invaders – with a lot of wit and supernatural strength. It is a magic potion that the Druid Getafix makes for Asterix and the other stubborn Gauls that boosts their brawn, but only if made with mistletoe harvested with a golden sickle.

Mistletoe is very capable of defending itself; its name, *Viscum album*, refers to the viscous and sticky insides of its fruits, a stringy slime that is – like all other parts of the plant – toxic to humans and that very few animals can eat without harm. The mistle thrush is one such creature. It returns the favour by excreting the semi-digested slime together with the seeds, which stick to the branches of a tree, depending on the bird's accuracy, of course. If the tree is a suitable host, the seeds sprout. The mistle thrush plays an active part in the spread of the parasite; in the past, however, it also contributed to its own demise as the sticky berries of the mistletoe were used to make birdlime, used to trap and capture mistle thrushes and other birds.

Humans aren't the only ones to hunt with sticky substances, and birds aren't the only ones to find themselves in predators' clutches. Hunting with glue works in every medium, but while prey in aquatic environments might swim or float by the trap, terrestrial predators often have to make their slime meet its victim. The oceans, as we have seen, are rich in nutritious particles and passive plankton, floating into the jaws and mucous traps of predators or steered in the right direction by specially generated micro-currents. Air, by contrast, offers little in the way of food, although tiny life forms can also be found riding the winds of their own accord. Yet most of these airborne organisms don't drift, but could fly around potential traps. This is why many spiders weave invisible webs laced with sticky slime across regularly frequented corridors in the air.

But this is a game of chance, while other predators actively lure prey to their doom. In a forest in Australia, I find gnat larvae of the species *Arachnocampa luminosa* crawling along at eye level through long tubes of slime on an overgrown slope. It is hard to believe, but these tiny predators spin a deadly slime trap. They put out a mess of dangling fishing lines dotted with sticky slime droplets that gleam like strings of translucent pearls. During the day they might be mistaken for dewdrops, but at night they can't be seen at all. Instead the larvae's bodies will light up, creating bright dots, an irresistible fake firmament of bluish lights for flying insects that will find themselves entangled in the sticky threads on their way to these treacherous stars.

Other terrestrial trappers prefer a more minimalist approach – and bomb their prey with one huge slime trap. The bolas spider launches a single strand of silk at its prey; one end bears a sticky drop of secretion. It works like a soft version of a medieval morning star, striking its victim without warning, leaving no chance of escape. Velvet worms take this long-distance hunting even further. Chasing after dinner on their short and stubby legs is out of the question, but why would they even try, with two large glands in their heads that shoot streams of slime? The sticky secretion reaches astonishingly far at astonishingly high pressure and hardens fast to immobilize even prey like larger insects, spiders and snails. Like any good weapon, this cannon provides defence as well by incapacitating any hungry – and not too big or strong – predator foolish enough to pursue a velvet worm.

As we have seen, frogs and other amphibians also use slime to catch their prey, thanks to their viscoelastic saliva. Some reptiles also know the trick. The Texas horned lizard (*Phrynosoma cornutum*) enjoys such a dangerous diet that it even requires twice the sticky secretions. This little lizard's dish of choice is the aggressive harvester ant, which can give a nasty nip and has a poisonous sting. But even when its sticky tongue has reeled them in, it's quite another feat to swallow the ants safely. The lizard packs them inside a thick ball of mucus which secures the ants' jaws, stings and entire bodies, making it easier to gulp them down.

But sticky tongues can do a whole lot more; woodpeckers, for instance, use them as a multifaceted tool to hunt. It had long been thought that these birds hack little holes in trees to harpoon the insects hidden inside with their tongues, the tips of which are hard and equipped with barbs. But this myth was debunked when the German film-maker Heinz Sielmann had the birds star in one of his pioneering nature films in the 1950s. For the first time, his footage showed them inside their nest and hunting for prey in rotting trees. Sielmann became known in the UK as 'Mr Woodpecker' thanks to the internationally successful documentary, which helped reveal some of the birds' secrets at the time and one more only a few years ago: 'Re-analysis of Sielmann's footage, however, showed that larvae are *not* pierced, but simply adhere to the sticky saliva at the end of the tongue,' writes the ornithologist Tim Birkhead.

A study of the Guadeloupe woodpecker from the Lesser Antilles confirmed the behaviour and revealed that the birds' tongues are rich in touch sensors at the tips – the better to tell, either by touch or taste, when they have made contact with a prey item. 'In turn, the insect larva was hardly passive on sensing the woodpecker's tongue, and either retreated or grasped on to the sides of its hole with its legs, making it difficult for the woodpecker to dislodge it,' writes Birkhead. 'Through a combination of sticky saliva, a barbed surface and an extraordinarily prehensile tip – but no piercing – the Guadeloupe woodpecker was able to extract its reluctant prey.'

In other cases, saliva even manages to keep pathogens down. Holy fire, or *Ignis sacer*, was a common ailment in the Middle Ages, causing, among other things, delusions and severe tissue damage. The Antonines, monks of a Roman Catholic congregation, established hospitals throughout Europe for the afflicted, whose suffering could be neither prevented nor treated at the time. We know now that it was caused by the ergot fungus, which infects grains and grasses but whose toxins protect the affected plants and prevent them from being eaten. It is now a rare condition among humans because infected cereal is disposed of before it reaches the food chain. Elk (*Alces alces*), however, have to look out for themselves – by drooling. Their saliva is a counter-weapon containing substances which inhibit fungal growth. It's a case of 'the more the merrier':

the animals graze mainly the same areas, drooling repeatedly on undergrowth, such that dangerous pathogens receive frequent doses of saliva as a suppressant.

A whole host of different animals use their saliva as a weapon and some of them pose a risk to humans too. Take the case of Grace Olive Wiley, the American herpetologist, whose long relationship with venomous snakes ended in catastrophe. On 20 July 1948, Wiley's collection of snakes was to be photographed and she posed, as she often did, with an Indian cobra. On this occasion, however, the snake was startled by the camera flash and bit Wiley on the finger. Wiley had nurtured a close relationship with all kinds of dangerous reptiles over several decades and succeeded in making a name for herself as a scientist. In 1928 she managed to breed rattlesnakes in captivity for the first time. Even deadly poisonous snakes could be tamed, in her view, and she allowed them to move around her home freely. She doesn't appear to have had any close shaves until the day of that fateful interview, and even propagated her lifestyle. 'The rattlesnake, once you get to know him, is a lovable creature,' she told *The Milwaukee Sentinel* on 22 September 1929.

Snake venom is designed to incapacitate prey and adversaries alike; it's a cocktail of toxins which cause paralysis, haemorrhaging and tissue damage. All snakes are likely descended from one poisonous reptilian ancestor, so even constrictors like pythons exhibit toxins in small quantities. However, their venom glands were adapted to produce more slime instead, which acts as a lubricant, helping them to swallow feathery and hairy prey without a problem. Their genuinely venomous relatives are furnished with a similar tool, boasting highly specialized salivary glands in their heads which produce the toxins, store them in a kind of chamber and, once they're needed, squeeze them through a passageway towards the snakes' hollow teeth, which inject the toxins into their victims.

In vipers and elapid snakes, this passageway leads over another gland with an unknown function. It appears to produce venom in the jararaca snake (*Bothrops jararaca*) though, possibly as an emergency supply when the larger reservoir runs dry. There is another speculation that its secretions may activate the stabilized venom on the way to the hollow teeth. This is supported by the fact that large cells at the gland's opening produce lots of slime, possibly protecting the tissue from its own toxins. Surprisingly, we find a similar set-up in our own bodies, where it's not toxins but corrosives that need to be kept in check. In a mammal's stomach, glands produce acid and digestive enzymes, but protect their own openings with thick layers of slime.

To date, mucin-like toxins have only been found in the venom of the cobra, though they do not form a viscous gel. They are part of a highly potent cocktail, one which, two hours after the fateful bite, ultimately cost Grace Olive Wiley her life. Other than as a buffer against self-harm, slime does not appear to play much of a role, if any, in the production of snake venom. Unless, of course, it is the target. 'Goo-eating snakes' are species that specialize in preying on invertebrates like worms and snails, which have to be ripped from their shells first and present another challenge right after: how do you handle highly flexible and viscous prey that is dripping in mucus? These snakes secrete proteins in their mouth that were long thought to be venomous but might actually somehow help to control the mucus – and to get those slippery snails down.

Slime doesn't always have to be a deterrent, though. Most terrestrial snails are hermaphroditic, both male and female, making them capable of asexual reproduction. But still, they prefer to mate in an often elaborate – and not necessarily vanilla – slimy ritual. One version of this sees the partners stabbing one another with sharp 'love darts'. The snail *Euhadra subnimbosa* of Japan even goes so far in its 'martial courting' as to ram these calcareous needles thousands of times into its mate's body, as the evolutionary biologist Menno Schilthuizen at the Naturalis Biodiversity Center in Leiden writes. Some scientists call this 'hormone warfare', since at the heart of the ritual is stringy slime covering the love darts and laden with substances that dramatically improve the shooter's chances of reproductive success. Inside the partner's body the slime helps the release of the shooter's sperm, which will otherwise be quickly digested before it has a chance to reach the eggs.

Love darts are certainly unusual, but the mating ritual of the leopard slug (*Limax maximus*) is even more extraordinary. The two partners abseil down from an overhanging branch, entwined in a never-ending spiral and entangled in an elastic thread of slime like a romantic tightrope act. As an encore, both slugs will – helped by gravity – extend their bluish penises. From openings in their heads. These organs can measure several times the length of their bodies – even up to 90cm in one species – and will also wrap around one another, maybe to douse each other in hormones, without violence this time. Each penis transports a sperm sac to its tip, where an exchange takes place in the end. Oh, and once they're done, one of the slugs gets to eat the now abandoned slime trapeze.

Even if they skipped the meal, ants would probably take the chance to dine out on the slime. They go wild over snails' secretions, inadvertently playing a game of parasitic roulette each time they dig in, as many dangerous foes might lurk there. The lancet liver fluke (*Dicrocoelium dendriticum*), for example, infests sheep and goats, but only at the end of a complex life cycle. The first step requires a snail to pick up the eggs from the excrement of a grazing animal. Inside the mollusc's body the larvae will hatch and transform into a resilient gel-coated cyst which in turn is excreted by the snail. This slime is probably enriched with enticing pheromones which makes it an irresistible treat for ants. Inside the insect's body, the parasite continues to grow, seizing control by impacting the ant's nervous system. Like little more than a zombie, the ant will deliberately put itself in harm's way by climbing up a stalk of grass and holding fast with its mouthparts, soon to be munched down by a grazing sheep or goat.

Other parasites hunt for slime themselves and feed off it. Tongue worms are actually parasitic arthropods but have extremely simplified bodies devoid of any crustacean characteristics. They live in their hosts' nasal cavities, as does *Linguatula arctica*, which exclusively afflicts reindeer and eats their slimy secretions. Yet a pronounced penchant for slime and other excretions tends to be the exception. Few animals will actively seek them out and one insect makes a defensive virtue of this general revulsion: the cereal leaf beetle, with its system of self-defence. The insect is a much-feared pest, affecting oats, rye, wheat and many other crops. Its larvae are particularly devastating, but they are well protected against predators. Their glistening shells are made of their own excrement, embedded in a slimy matrix, arching over their bodies like a dome.

21

Roots, Shoots and Leaves

In an ecstasy of eagerness for the food so near them,
the leaves turned upon each other. Two meeting
would suck together face to face . . . writhing
like some green worm, and at last, faint with the
violence of the paroxysm, would slowly separate,
falling apart as leeches gorged drop off the limbs.
A sticky dew glistened in the dimples, welled over,
and trickled down the leaf.

Philip Stewart Robinson, *Under the Sun*

The helpless creature is caught in the monster's clutches, slowly encircled by its elegant tentacles and swallowed whole. Dying slowly, the victim, a water flea, can for quite some time be seen moving inside the body of its killer. It is a polyp with tentacles: transparent, almost bodiless, just a spectre, as Professor Bulwer explains to his astonished students, who are observing the devious attack in a scene from *Nosferatu: A Symphony of Horror*, released in 1922. In this silent film, a classic of Weimar expressionism, Friedrich Wilhelm Murnau followed in the wake of Bram Stoker's *Dracula*. However, he also included short sequences from nature films of spiders, devilish freshwater polyps and other meat-eaters.

These include carnivorous plants, which the Canadian scientist Janet Janzen considers in the context of the movie a metaphor for the sense of social destabilization and loss of traditional boundaries in the cinema of the Weimar era. These 'demonic' plants are intended to show the bloodsucking count, Bulwer's real target, as one predator among many in nature. Another scene depicts a fly in the clutches of a carnivorous Venus flytrap. 'Astonishing isn't it, gentlemen?' reads the text, conveying the professor's words, 'Almost like a vampire.' Despite Murnau's attempt to rehabilitate the image of vampires, they retain a whiff of abnormality; even the undead can't quite get away with drinking human blood.

In Victorian times, even some plants fell into disrepute, not because they were to be found emerging pale and sharp-toothed from the compost heap, but because they nursed an unnatural hunger in life. Carnivorous greenery represented anarchy in the God-given order of nature. Such plants were viewed as zoophytes, peculiar hybrids which were neither plant alone nor animal. And if small, fly-catching plants managed to set nerves aflutter, stories of their bigger and hungrier relatives were even more blood-curdlingly thrilling. In April 1874, the *New York World* released a report by the 'German explorer' Carl Liche about an expedition into the jungles of Madagascar. Under the guidance of the 'savage but knowledgeable' Mkodo tribe, he claimed to have come across a plant nearly three metres tall and resembling the exotic pineapple, with equally long leaves and palps, which twined themselves round the plant like the snakes on Medusa's head.

A young Mkodo woman was made to drink the plant's viscous nectar, and was then wrapped in its supple tendrils and strangled. A terrible shrieking, a horrible laugh, a gurgling groan . . . and then plant sap, mixed with the victim's blood, ran down the plant's stem and was drunk by the Mkodo in 'an indescribable orgy'. The fabricated tale of the Mkodo, the man-eating plant and even the persona of Carl Liche could be traced back to one newspaper editor, who had printed plenty of other tall tales. Charles Darwin was working on a publication about carnivorous plants at the time and was not to be fooled.

Yet some other enthusiastic readers fell for this horror story and offshoots of *Crinoida dajeena*, the so-called Man-Eating Tree of

Madagascar, still haunt the internet – like the Nicaraguan Blood-sucking Vine. Audrey II from the film and musical *Little Shop of Horrors* eats people and incites murder in the hope of taking over the Earth – which is what you'd expect from a cosmic horror based loosely on the works of H.P. Lovecraft. In reality, the carnivorous plants which do exist on this planet are a little smaller and are simply trying to survive, mostly on poor soil and often with help from slime.

★★★

'Have you lost a contact lens?' a concerned walker pipes up on spotting me bent over a patch of French marshland, scanning the ground. He has no idea that I'm on the hunt for a monster, hoping to catch sight of some sundew, *Drosera*. This little carnivore, mere centimetres tall, should be hiding somewhere in the grassy scrub and, eventually, I find it. The tips of the delicate plant's leaves are broad, curved and covered in red protuberances. Each of these little hairs appears to be laden with a glistening drop of dew on its tip. It's actually a sticky kind of mucilage, as plant slimes often are called, in this case a highly specialized hydrogel trap. Insects or spiders get stuck to it before being wrapped up in the leaf and broken down by digestive secretions.

These sparse little plants seem to cling to life by carving out an existence for themselves with great effort in a habitat that offers sunlight and water but scarce nutrients. Almost all green carnivores must supplement the diet available from the ground with meat to ensure that, more than anything else, nitrogen comes flying their way. But it's not always just the hardy survivors that eke out an existence the carnivorous way. Years later in Australia, I find ground covered with thick carpets of *Drosera*, a battlefield littered with its victims' wings and other unpalatable remains. But where there is a slime, there is in all likelihood a goo-eater: the caterpillar of the plume moth *Trichoptilus parvulus* lives a dangerous life among the traps but neutralizes them by licking off the mucilage in order to eat the plant's leaves safely.

Other potential prey gets away by sheer strength, overcoming the sticky slime. This might be why some descendants of *Drosera* have evolved more sophisticated mechanisms. Professor Bulwer's Venus flytrap, *Dionaea muscipula*, for example sports a snap-jaw trap comprised of two halves with a central hinge, the edges of which are covered in stiff barbs. Only once the plant has sensed repeated movement will it close like a gap-toothed mouth, trapping its insect prey behind bars, no mucilage required. But even some slime traps have learnt to stop prey in its tracks effectively – with more slime. Butterworts, for example, cover their large leaves with an already sticky secretion but will smother any stuck insect in extra glue, giving it no chance to get away.

We humans are not above using slime to gain an advantage over our prey. Well into the nineteenth century, bird-catching was a widespread profession, and mistletoe and holly were two of the plants used to line the sticky traps: 'The term birdlime denotes a green, tough, stringy, very sticky, dough-like substance that melts under heat, is insoluble in water, and is therefore significantly different from glue, and is mixed with a little fat and used to coat the rods with which one catches birds,' wrote Johann Carl Leuchs in 1842 in *Instructions on the Preparation of Joiner's Glue, Isinglass Bone Jelly, Bird Lime and Solid Broth*.

After being deemed 'traditional' enough to be protected in some places, the practice of glue-trapping has only recently been banned in the European Union. In the past, professional bird-catchers could make their living by selling the birds for their beautiful songs or their tender meat. But what's *Pisonia*'s excuse? The tropical 'birdcatcher tree' is something of a botanical mystery. In order to spread, the plant employs the common tactic of allowing its slimy and barbed seeds to stick to the feathers of seabirds and be carried along. Young animals, however, are often weighed down by the pods, unable ever to get up again, and their bodies litter the area around the plant's roots and become caught in its crown. What remains a mystery, however, is whether this massacre benefits the plant in some way.

There's yet to be a clear definition of carnivory in plants, but five criteria are usually required: animals must be lured in, captured, killed, broken down by the plant's enzymes and then absorbed by the plant to its own advantage. Shepherd's purse (*Capsella bursa-pastoris*) is an unassuming plant that nonetheless goes quite far on this front. Its seeds lure in its victims, over 90 per cent of which do not survive the encounter, because the plant's slime glues the insect's mouth parts together, causing it to suffocate if it has not already died of exhaustion. Shepherd's purse was long considered a protocarnivore – a former or nascent carnivore – because it could not be proved that the plant exploited its prey for its own ends. However, new studies have shown that killing nematodes helps its seeds germinate more efficiently, and its seedlings to grow particularly lush. It is possible that carnivory in plants is not a clear-cut characteristic but part of a spectrum of options to take up the nutrients they require.

If sticky slime itself predisposes plants towards some sort of carnivory, an army of future meat-eaters could be waiting in the wings. Mucilages have been found in shrubs and flowering plants, but also in algae, lichen, mosses and ferns. They're secreted by roots, shoots, leaves and flowers – with a myriad of functions. A gel coating is particularly common and of considerable importance in the diaspores, the seeds and fruit, surrounding them, at least when they come into contact with water, in a jelly-like capsule. This is useful, for instance, in fruits which have to travel through a bat's digestive tract in order for the seeds to germinate. A slimy coating can ensure frictionless passage, like linseeds soaked in water, which are known for getting the sluggish bowels of constipated humans moving.

A diaspore's gel coating provides a classic barrier at its interface with its environment, protecting it and creating a pleasant microclimate. Its numerous additional uses will depend on what is needed, even helping with safe travel beyond an animal's guts. If diaspores land in a brook or river, for example, their slimy coating enables them to float to the surface, which gives them a better chance of survival and a landing on dry land again. But water is not the only option for long-distance travel. If seeds are coated in a strong enough slimy adhesive they can stick to birds or other animals as stowaways with a chance to conquer new habitats.

In dry regions and deserts, but in temperate areas as well, mucilage can conserve water to protect diaspores from drying out. At maximum water absorption, the seeds of *Artemisia sphaerocephala*, for example, swell to almost 600 times their original size. Adhesion is another important job, especially for seeds that aren't keen on travel. Some would end up as food for ants, never to be seen again, if they were not immediately cemented to the ground with a little help from their slimes. This is often reusable glue that can dry out and become adhesive again. But even the drying-out can have a purpose: when sand or other particles get stuck and coat the seed in a thick crust it is particularly bothersome to carry away, even for stubborn ants.

In an even more sophisticated twist, mucilage could support a plant's development. The gel coating on the seeds of sweet rocket (*Hesperis matronalis*) is shaped like a lens and focuses the light on a specific area of the seed, facilitating germination and activating important enzymes. It then helps the first roots in penetrating the soil, and may well also harbour a specific microbiota. Plants rely on this colony of microbes. Like all metazoan organisms, they are holobionts and nurture an intimate relationship with their resident microbes, giving them a choice of habitats. The most important and most active of these is the rhizosphere, or the thin layer of soil around roots.

The term was coined by the German agronomist Lorenz Hiltner in 1904 for soil influenced by plant roots. But modern science sees the rhizosphere reaching wider than just a small volume of soil. It is a dynamic system of interacting processes between components like the ever-shifting microbial community, invertebrates, water, soil, nutrients and root products like mucilage, which here is often called mucigel. And plants are not only passive partners at that essential yet vulnerable interface. After all, they are unable to escape harsh conditions like drought, over-salinization or temperature fluctuations: they need to shape their environment actively.

But the rhizosphere is not only a plant's most important interface; it is also an ecological hotspot. Roots and the surrounding soil are considered one of the most complex ecosystems on Earth. Bacteria, fungi, archaea and other microbes thrive here and might be inclined to co-operate with the plant if they're offered an incentive to settle. Root exudates are the currency they trade in. These are secreted metabolites, like amino acids and sugars, but also chemical compounds to keep pathogens away – and slimy mucilage. The choice of exudates will, for example, shape the microbial community in the rhizosphere, which will in turn influence which exudates they get served, much like the interplay between our gut, our diet and our microbiota. If all goes well, root-associated microbes help to keep pathogens away and support plant growth.

We still know relatively little about the structure and regulation of the rhizosphere, even though the microbiota influences a plant's productivity and immunity, its capacity to withstand environmental stress, the health of the soil, and, in so doing, the foundations of our diets. Its genetic material is even considered a 'second genome' for plants, which can access this reservoir and the capabilities it brings when needed. This is also why the rhizosphere's microbiota has become a focus of research. In the future, it is hoped that tailored microbial mixes will be used to toughen up plants or restore tired soils. This targeted colonization by microbes could help to open up new arable land and counteract the effects of climate change on agriculture.

However, this habitat is difficult to access and the rhizosphere is highly dynamic. The composition of the microbiota is actively regulated by the plant and varies according to species and age, soil conditions, environment and current general conditions, as well as other factors. In evolutionary terms, it is an ancient form of co-operation which probably began at least 460 million years ago, perhaps earlier, when the land was colonized by plants. One challenge is the provision of nitrogen, an essential nutrient, which is present in abundance but cannot be used by plants in the form in which it naturally occurs. It is for this reason that legumes such as beans and peas have what are known as rhizobiaceae, or root-nodule bacteria, on their roots which fix nitrogen in the soil, rendering it usable by plants.

For corn and other crops, however, artificial fertilizers are used. These require enormous energy to produce and are partly washed away into bodies of water which cannot cope with an oversupply of nutrients. A new solution is needed – and it might come in the form of a giant Mexican corn. This plant takes up nitrogen directly from the air via its thick, red aerial roots, which encircle the stem and are covered in a thick slime, an extremely sweet hydrogel, which drips on to the ground below in gleaming droplets. A complex microbiota makes its home here, fixing nitrogen which it supplies to the corn. If this principle could be applied to other plants, it would make nitrogen-poor soil viable as arable land and reduce the use of fertilizer. But this is not much more than an idea so far, and not a solution we are ever likely to be able to count on.

Easier to grasp might be the functions of mucigel, the slimy substance root tips secrete which influences physical properties of the soil. Mucigel stores water and keeps it close to the roots while connecting them to the soil by stabilizing it as a viscous glue. Some scientists even speak of a 'rhizosheath', composed of binding material like mucigel and adhering soil. It also lubricates the root tips' way into the soil – a risky endeavour because they will penetrate heavily populated, uncharted microbial territory.

It seems likely that mucigel offers plants at least some protection, just as mucus shields our inner surfaces. Each sensitive root tip is covered in loose boundary cells like little tiles, which gradually slip off and survive for weeks or even months in the rhizosphere. They play an active role in defence, swarming if danger is imminent. In the event of an attack, root tips will crank up all these defences, accelerating production of mucilage and of boundary cells that produce a matrix of sticky DNA to immobilize pathogens. These kinds of extracellular traps are common, an 'ancient weapon of multiple kingdoms' as a recent publication has it. They have been found as a defensive mechanism in plants, but also in invertebrates like mussels and slugs and in vertebrates including mammals.

Our neutrophils, a type of white blood cell, behave in a similar way, ensnaring pathogens in sticky DNA nets or even parasitic hookworm larvae as soon as they penetrate our skin. In the soil, however, another deadly web may come alive. Akin to the roots of plants, fungi build underground networks of thread-like hyphen. That mycelium is coated in slime, a perfect trap for some species. If the mucilage is sticky enough, even tiny worms become stuck before being impaled, paralysed and broken down by the fungus hyphen. Even more original, perhaps, is the lasso, a ring of mycelium which swells quickly as soon as a nematode creeps through it. These mechanisms can also be activated by certain bacteria the nematodes eat. The enemy of their enemy thus becomes their friend, and in the future we may well be able to control parasitic nematodes and pests with the help of fungus – which will be all too happy to lend a hand. It's not about self-defence for the trap-setters: these fungi are carnivorous killers.

VII
Environment

Slime is a substance of interfaces, including those between the spheres of air, water and earth. It often remains hidden, but it affects every interaction that takes place. Climate change could have profound consequences on these processes, with slime production increasing in our warming, acidifying oceans. If the balance is disturbed, the effects may be felt worldwide, impacting the ability of our oceans to act as carbon stores and counteract climate change. The movements of the tectonic plates, the planet's floating land masses, are also influenced by slime deposited as marine sediment. And many ecosystems are being dramatically depleted. In the long term, this could transform the face of the Earth as we know it. If the empty oceans become dominated by algal or jellyfish blooms and microbial mats, it would be a man-made backwards step in the story of evolution: a new era of slime.

22

Gaia and the Gel

The great Pacific tide was coming in and every few seconds the relatively still water of the lagoon heaved forward an inch. There were creatures that lived in this last fling of the sea, tiny transparencies that came questing in with the water over the hot dry sand . . . Like a myriad of tiny teeth in a saw, the transparencies came scavenging over the beach.
William Golding, *Lord of the Flies*

Ebble Thatch Cottage is the kind of place you'd expect to encounter in the pages of *The Hobbit*. Yet this little thatched house overgrown with roses, nestled in the village of Bowerchalke, was home to a man who owed his worldwide fame to a literary dystopia. William Golding was teaching at a boys' school in nearby Salisbury when he wrote *Lord of the Flies*, the story of a group of boys fighting for survival on a Pacific island following a plane crash. The mask of civilization soon falls away, conflicts are dealt with bloodily, and one child dies at the hand of another. Evil taints many scenes, from a pig's head swarming with flies to the seemingly harmless game in which little Henry tries to control the movements of the mysterious sea creatures he spies on the beach.

Golding received the Nobel Prize in Literature for this disturbing Robinsonade in 1983. On receiving the award, he gave a speech in which he spoke about a treasured childhood experience which may have inspired Henry's game on the beach. In Golding's memory, however, the sea was at a particularly low ebb. It was the school holidays and the retreating water revealed a cavern, a last recess before the even more mysterious deep sea, which had strange inhabitants he had found nowhere else. 'Only a hand's breadth away in the last few inches of still water they flowered, grey, green and purple, palpably alive.' Golding doesn't reveal what species these creatures were, and he probably never knew.

Perhaps this is where Rachel Carson can help; she writes in *The Edge of the Sea* of another pool in another cave on another coast, one which could only be seen briefly during the lowest tide of the year: 'Under water that was clear as glass the pool was carpeted with green sponge. Grey patches of sea squirts glistened on the ceiling and colonies of soft coral were a pale apricot colour.' Whenever she visited this magical zone of the low water of the spring tides, she writes, she was on the lookout for the most delicately beautiful of all the shore's inhabitants – flowers that are not plants but animals, blooming on the threshold of the deeper sea. These were hydroids of the genus *Tubularia*, creatures 'so exquisitely fashioned that they seemed unreal, their beauty too fragile to exist in a world of crushing force'.

Golding may have marvelled at similarly fragile creatures, but it's clear that their strangeness added to the wonder of this unexpected encounter, which moved the young boy so deeply that he returned to the same patch of coast as an adult, only to find that the cave had been abandoned: 'Nothing lives there any more. It is all very clean now, ironically so, clean sand, clean water, clean rock.' Whether it was a natural process, fuel oil, sewage or chemicals which destroyed that little bit of childhood magic and mystery, he cannot say. But it doesn't matter, as this was 'only one tiny example among millions' of how we impoverish the 'only planet that we have to live on'. William Golding commended to his audience, in Stockholm and around the world, the notion of our planet as 'Gaia Mater'. In Greek mythology, Gaia is the ancestral mother of all life; as the prefix 'geo' she is eternalized in all earth sciences.

Thanks to Golding, however, Gaia now represents a decades-long and at times bitter dispute. And it all came about during a walk with James E. Lovelock, a neighbour of Golding's from Bowerchalke and an unconventional thinker. Lovelock had spent time working as a scientist for NASA on a project on the search for extraterrestrial life. One afternoon in June 1965, however, he was considering his own planet. Wasn't the Earth itself a kind of superorganism? All life forms interact with the surface of the Earth, the oceans and the atmosphere. It was Lovelock's conviction that this phenomenon gave rise to conditions on Earth which were ever more habitable, in a sense, life encouraging the existence of life on Earth. He had the concept; all he needed was a catchy name for his self-regulating and self-supporting system. Golding suggested the name Gaia and the goddess herself was reborn as a theory.

The New Age movement immediately claimed the concept of a supposedly holistic Mother Earth for themselves and gave it esoteric foundations. It didn't help that Lovelock himself protested as vehemently against this co-opting of the theory as did Lynn Margulis, another creative thinker and highly influential scientist, who would go on to become his ally in all things Gaia. The hypothesis was disputed within the scientific community anyway, but the hippie connection certainly added another gust to the strong headwind of opposition. Criticism continues today, in part because proof of the ultimate hospitability of the Earth is still pending. Later, we will encounter a theory which asserts the opposite and is similarly named after a mythological mother, the murderous Medea. In both cases, however, it remains undisputed that the deep links between Earth and her inhabitants as well as their activities have always shaped the planet, even though we don't know nearly enough about it.

★★★

Life has had a formative influence on the planet from the beginning, even down to its very bones: microbes can tear down rock and rebuild it. They are essential to geochemical processes like the carbon and nitrogen cycles. They were the first to enrich the atmosphere and the oceans with oxygen, which allowed higher organisms to develop. They have always been the central biological cog in the workings of our planet, yet one we've caught only a glimpse of so far. One crucial factor in this unchallenged success was their early appearance, with not a competitor in sight for billions of years.

No doubt, life has shaped the planet. But it has been first and foremost microbial life which means that we all live in a primarily microbial world. During their long, undisputed reign, microbes were able to conquer all of the inhabitable niches on the planet, dividing up into an unimaginable richness of different species and, in doing so, establishing the biosphere. This is the Earth's organic shell, the entirety of life on Earth. The limits of the biosphere are defined by microbial capacity for survival. One last and presumably unconquerable frontier sits somewhere high above the clouds, the other is somewhere deep in the rock under the surface of the Earth, constituting what is known as the 'deep biosphere'.

It comprises all life in a vast area, starting a few metres below the planet's surface but extending at least five kilometres into the deep. Microbes have been found miles underground beneath the Rocky Mountains, under the seabed and in mines boasting almost intolerable conditions. 'At two miles beneath the surface of the Earth, the heat and the humidity were almost all I could think about,' writes the geomicrobiologist Tullis Onstott at Princeton University about his visit to a South African gold mine. His ordeals are worth it, though; the scientific community hopes that the depths will reveal clues about the origins and spread of life on our planet, and possibly on other planets too.

A few decades ago, life this far beneath the Earth's surface was completely unknown to us and it was hardly conceivable that microbes might be able to survive in the tiniest pores without air and light, under immense pressure and in highly acidic or alkaline conditions – trapped down there for thousands, millions, even billions of years. The Deep Carbon Observatory changed that view: like HMS *Challenger* charting the oceans in the second half of the nineteenth century, this global community of more than 1,000 scientists from all over the world embarked on a ten-year exploration of the terra incognita beneath our feet in 2009 to find out 'what's known, unknown and unknowable about Earth's most pristine ecosystem'.

Life on Earth is carbon-based, yet 90 per cent of the planet's carbon resides inside the planet, so how does that deep carbon impact the oceans, atmosphere and life on the surface? The project was supposed to begin to look for answers to this question and found a vast ecosystem of hundreds of millions of microbial life forms thriving beneath the planet's surface. The volume of the deep biosphere is now thought to be twice as large as all the oceans combined. It could comprise 70 per cent of all bacteria and unicellular archaea on the planet, with the biomass of these microbes potentially outnumbering that of humans nearly 400 times.

Yet the microbes' extreme existence has its price, and if they are not quite dead yet, they often are not truly alive either. In the rocky deep, a change of microbial generations can potentially take thousands of years, as opposed to the usual minutes. Other life processes may be slower than the organ piece in Halberstadt played many times over as well. And yet it is life which has adapted to a lack of access to sunlight or nutrients by using other sources of energy from its mineral surroundings, a morsel every few thousand years. Microbes' genetic equipment will help – and probably allow for slimy protection as well. In a recent study, scientists reported on drillings in rocks that lie deep beneath the ocean's crust and are up to 104 million years old. They found life there in astounding densities and huddling together in biofilms.

Microbes are the foundation and engineers of global ecosystems, as in the depth of the planet so on its surface, where no single habitat has been found free of them so far, from the hottest deserts to the Arctic and darkness of the deep sea. They have had billions of years to adapt to these extreme places and paved the way for evolutionary newcomers – basically, the rest of us. Other species benefit from the aeons-long microbial activities that tap into otherwise unusable sources of energy or buffer extremely inhospitable conditions, in short, prepare new ground for all kinds of life.

The focal points of these interactions are often where the global spheres of rock, earth, water and air come into contact with each other. These interfaces can be found in underground caves as well as on the planet's surface, in deserts and other dry areas, in the deep sea as well as on the ocean's surface. Microbes live here, settle here and shape these habitats. Consequently, almost every exchange between the planet's spheres involves microbial habitats, be it the transfer of a molecule of gas, a solid particle or the transfer of energy in the form of sunlight. But even microbes couldn't have conquered every hostile terrain and survived in some of the more extreme habitats without biofilms.

Slime is a microbe's protective armour, anchor and cement, invisibly holding whole landscapes together. The Earth would look very different without it even though the goo has only been inadequately researched in many cases. But this is changing, both in the deep and on the surface of the planet. The role of biological adhesives is becoming a major focus of research, and not only because they're under threat. Many of the ecosystems which slime has had a role in shaping could be fundamentally changed by environmental destruction, the loss of biodiversity and a warming climate. That is a threat to all of life, and humans too. Once-fertile areas could become dry and deserts expand, coastlines could rapidly erode and global cycles could be destabilized.

Here's a little thought-experiment to illustrate how widespread the slimy matrix is and how tightly interwoven it is with our planet and biosphere. If the Earth suddenly disappeared with all its life forms and all its water, for a few moments afterwards a slimy echo of its contours would remain. This is the gel that microbes and other species such as worms, corals and snails secrete in such large quantities that they actually change the characteristics of entire areas of land and marine habitats. Its gossamer outline would trace vast swathes of the Earth's surface, from caves to ocean waves frozen in motion, a delicate, slimy echo. In place of the sea there would be a sticky web set in the shape of lost coral reefs and the ocean floor.

Gaia notwithstanding, the Earth is not a superorganism. But we may be able to grasp its vulnerability more easily if we compare it to the human body. Our tissue and organs are composed of specialist cells and blood vessels, held together by a gel-like matrix to form a three-dimensional entity. Without this kind of slime we would disintegrate. Across the Earth, the different domains of air, water and solid matter form a three-dimensional patchwork with allocated roles. The sea, for example, acts like a gas-exchanger, taking in CO_2 and releasing oxygen. This exchange with the atmosphere occurs via a gel-like membrane on the surface of the oceans – reminiscent to a mucous membrane. Deserts and dry regions are also covered in a mucilaginous surface which works to counter erosion and evaporation, a bit like our skin.

Alongside Gaia, perhaps what we need is a narrative which captures the whole globe, which gets under its skin, looks to its past and to its future. The story must also present slime as a glue that holds vast stretches of our environment together and facilitates a fresh understanding of landscapes we had thought we knew well. But that delicate balance is under threat and changing already. Of one thing only we may be sure: microbes will adapt, might even in some places become more active and slimier in the process. But a surplus of gel could be just as problematic as a dearth of slime; whole habitats could irreversibly shift. We need to take a closer look at the seemingly unassuming interfaces of our world, the ocean's slimy skin, the gel of coral reefs or the sticky marine snow transporting carbon into the deep – and countering climate change. They might seem unsightly at first, but slimes, and those who produce them, bind our world together.

23

The Skin of the Sea

> With the float or sail set diagonally across
> its base, the creature sails before the wind;
> looking down into the clear water one can see
> the tentacles trailing far below the float. The
> Portuguese man-of-war is like a small fishing
> boat trailing a drift net, but its 'net' is more
> nearly like a group of high-voltage wires, so
> deadly is the sting of the tentacles . . .
>
> Rachel Carson, *The Edge of the Sea*

Fifty-six long days on the open ocean. Fifty-six days of storms, sea sickness and scurvy before land came into view on 9 November 1620, promising religious freedom. But the *Mayflower* had sailed off-course, travelling further north than intended all the way to Cape Cod, a peninsula near Boston. Should they sail on, or disembark? The exhausted seafarers opted for dry land, and today a slim monument in Provincetown, Massachusetts commemorates the arrival of the English pilgrims on American soil. Economic considerations also played a part in their decision, of course, and Cape Cod provided rich fishing grounds for centuries. Whale-hunting alone offered meat, oil for lamps and trains, as well as baleen plates for corsets. In *The Perfect Storm*, Sebastian Junger writes of men working at sea in the waters around the coast of New England, explaining that, in years gone by, 'fishing in Gloucester was the roughest sort of business, and one of the deadliest'. Little has changed.

Hic sunt dracones. Junger quotes Sir Walter Scott, whose words still ring true today: 'It's no fish ye're buying, it's men's lives.' The surface of the sea had always concealed beneath it the promise of new realms, food and riches, but monsters also lurked in its murky depths. In the Middle Ages, illustrators could earn more if they decorated their nautical maps with plenty of mysterious creatures. We're familiar with that same frisson of horror from the big screen, when the dorsal fin of a Great White slices through the deceptively still surface of the sea, or when Godzilla emerges from the flood. The ocean's surface is the thin veil that will hide the monsters of our psyche, as H.P. Lovecraft surely knew.

But not all the forces lurking beneath the 'anonymous surface of the sea', as the historian Nathaniel Philbrick calls it, can be controlled. There are monsters, and some are of our own making. The *Essex*, an American whaler, sank in the South Pacific in 1820 after being rammed by an angry sperm whale that decided to fight back. Twenty men abandoned ship and sought refuge in three little boats; eight survived, only by eating the bodies of their dead friends – which made the *Essex* as well known as the *Titanic* is today. For one man the attack itself proved a source of inspiration, as Philbrick explains. Herman Melville's Captain Ahab sets out from Nantucket, an island off Cape Cod, to scour the ocean for Moby Dick. Melville keeps his eye on the ocean's surface membrane and the world concealed beneath it. As his narrator, Ishmael, says: 'These are the times of dreamy quietude, when beholding the tranquil beauty and brilliancy of the ocean's skin, one forgets the tiger heart that pants beneath it; and would not willingly remember, that this velvet paw but conceals a remorseless fang.'

★★★

The skin of the sea has long been viewed as little more than a boundary between the human world and the deadly wonders of the ocean. But it is also a vast interface between the sea and the atmosphere, overlapping at its edges with dry land, and covering 70 per cent of the surface of our planet. And the skin of the sea is composed of a delicate layer of gel just a fraction of a millimetre thick. Is this where the sperm of Uranus has been sloshing around since the time of myth? Uranus was Gaia's first son and fathered many children with her. One of these was Kronos, who – at Gaia's behest – castrated Uranus. He cut off his testicles with a sickle and threw them into the sea, where Uranus's semen produced foam out of which rose Aphrodite, the goddess of love.

Science prefers to think of this physical boundary as the gel-like 'sea-surface microlayer' (SML). There are similar surface layers in freshwater ecosystems. The skin of the sea boasts unique chemical and biological properties and is clearly differentiated from the layers of water beneath it. It is fed by organic particles, such as slimes from different marine species, and proteins and fats which rise to the surface from the water column. These condense to form a carpet of nutrients, which is broken down by microbes that colonize the surface and form slimy biofilms – or at least lend the layer 'biofilm-like traits', according to Oliver Wurl from the University of Oldenburg, who is researching this gossamer-thin interface and how it may change in the future.

At least a third of the climate-relevant greenhouse gas CO_2, which humans have produced since the Industrial Revolution, has been dissolved into the ocean, Wurl states in a recent publication with colleagues. 'Consequently, understanding how the ocean absorbs CO_2 is critical for the prediction of climate change.' And since the SML is a ubiquitous diffusion layer between the ocean and the atmosphere, it is 'a primary point for the air–sea exchange of greenhouse gases . . . heat and particles'. The sea's membrane has yet to undergo extensive research; it is too vast, too thin and too changeable to allow for comprehensive analysis. And it is not uniformly built: excessive amounts of organic matter accumulate to so-called slicks on the sea's surface that can potentially reduce the transfer of climate-relevant gases.

It is clear that climate change leads to a warming of the sea and its acidification, which may in turn change the sea-surface microlayer in a fundamental way. 'If the ocean warms up, this could cause phytoplankton to become more active and the gel layer would grow thicker,' says Wurl. 'This would probably also affect the exchange between the atmosphere and the ocean.' But how? Transfer could become easier and quicker – or be slowed down. Or some transfers might accelerate, while others change in any number of ways.

This uncertainty about what is to come concerns not only the exchange of gases like CO_2 that needs to be taken up from the atmosphere, or oxygen that is being released from the ocean. The SML covers roughly two-thirds of the planet's surface and plays many roles beyond that of a diffusion layer: if bubbles from the ocean burst at the surface, for example, they spray organic material and bacteria neatly packaged in microscopic saltwater droplets into the air. These aerosols can be carried up into the atmosphere and contribute to cloud formation. But this is another mysterious process in flux; as a recent publication states, 'Marine aerosol-cloud interactions represent one of the largest uncertainties in understanding climate change.'

Alongside the sea floor and the open ocean, the gel layer constitutes its own domain. It is home to dense populations of different microbes. Viruses are one more central cog with a likely ecological impact on the other microbes and air–water exchanges about which we know very little. They 'remain the most enigmatic biological entities in boundary surface layers', as one publication has it. But they are only part of a unique ecosystem in these challenging conditions, alongside strong winds, powerful waves and extreme solar radiation with wild fluctuations in temperature and salinity – the neuston. Microbes, molluscs and some invertebrates, fish, larvae and fry all live in, at or on the skin of the sea and must tolerate its exposed nature.

Sea skaters of the genus *Halobates* are widespread but they are a curiosity in the world of insects, for whom aquatic and especially marine habitats are mostly foreign. *Halobates* do little more than dip their toes in the sea; tiny water-repelling hairs on their legs allow them to glide over the surface of the water, looking for prey to suck dry, such as floating eggs and larvae. A sea skater is an out-and-out Narcissus, spending its life gazing at its own reflection. On occasion, I imagine, it might spot another face floating beneath it, daubed in blue war paint. This is the small sea slug *Glaucus atlanticus*; its gaze is fixed skywards as it glides along under the skin of the sea, ignorant to any and all who might emerge beneath it from the watery depths. In Greek mythology, a magical herb turned the fisherman Glaucus into a sea god with fins instead of fingers. Fittingly, these sea slugs grow long cerata, finger-shaped outgrowths which make this mollusc look as if it's floating lazily in the water.

I'd like to offer another interpretation: Glauce, also known as Creusa, was the unfortunate daughter of King Creon, who received a poisoned dress from the jealous Medea. *Glaucus*, the dragon slug, on the other hand, makes do with deadly gloves. It preys on floating *Gelata*, particularly the Portuguese man o'war, whose toxic nematocysts are highly effective weapons, as Rachel Carson so aptly described. But they don't deter all predators: the slug is happy to use the nematocysts for its own ends. It's a risky business; the toxin-loaded capsules can be triggered by the slightest touch. This is not unlike the tricky handling of technical explosives that go off at the slightest tremor. The Swedish chemist Alfred Nobel decided to coat his newly invented dynamite in a kind of gel. Even today, so-called blasting gelatine – or gelignite – is used to level buildings without posing great risk to workers.

The *Glaucus* probably follows the same principle, coating the nematocysts it has captured in slime, enabling it to transport them intact through its own tissue towards the cerata. Accidents could be cushioned by chitinous plates which cover the creature's digestive tract and serve as armour. When predators attack, the nematocysts protect the sea slug, and since it accumulates the arsenal of more than one siphonophore its toxic effect can even surpass that of a single Portuguese man o'war. The thieving *Glaucus* is a pirate, as befits a member of the 'blue fleet'.

This term was coined by the British naturalist Alister Hardy for all the floating and gliding neustonic creatures which share their colour with the sea. He understood this colour to be protection against UV radiation, though it is now thought to be camouflage, an idea which Hardy may well have had himself. He worked as a camoufleur during the First World War, designing military camouflage – while sailing under false colours himself. As the author Peter Forbes writes, Hardy was only chosen for the role because he was confused with a professional artist of the same name. Camoufleurs faced a new challenge: for the first time their adversaries could scout an area by plane. How should large amounts of equipment and numbers of troops be concealed from prying eyes which could now take to the air?

An art revolution in the first decade of the twentieth century may have helped here, as the outlines of objects and figures were abolished. Cubism offered an extreme interpretation of this. As Forbes writes, Pablo Picasso and Georges Braque's multifaceted guitars looked almost like camouflage, a fact unlikely to have escaped the painters. The military knew that large, moving objects could not be hidden completely, but without clear outlines their size and speed became more difficult to assess. Forbes quotes an interview with Braque in which he insists that cubism wasn't the only style the army used for its camouflage: 'Before cubism we had impressionism, and the army used pale-blue uniforms, horizon-blue, atmospheric camouflage.' It's the colour of the sea as well, and may help *Glaucus* and the rest of Hardy's oceanic fleet to become hard to spot by airborne predators. But the sea slug's deceit extends in every direction; its pale back blends in against the sky when viewed from the ocean's depths.

Over half a century ago, Yuvenaly P. Zaitsev was investigating the neuston and the origins of its inhabitants. According to the Ukrainian scientist, these creatures developed from species based on the sea floor. It occurred to him that neustonic species had counterparts in the depths, forming a kind of mirror world. Down on the sea floor, snails crawl and crabs scuttle, while mussels and sea anemones are firmly anchored to a base. When searching for food, they graze the seabed or fish particles out of the open water. All these species have an equivalent on the water's surface – though they're often upside down. The predatory violet sea snail is part of Hardy's 'blue fleet'. From its foot, the sea snail produces a float made of slime which quickly hardens, enabling it to anchor itself to the surface and hang upside down in the open water. Neustonic crabs, on the other hand, must use their powerful paddle-feet to steer themselves around, while sea anemones have developed a pedal disc, and buoy barnacles a polystyrene-like float from which they hang, filtering prey out of the seawater.

All these creatures are adapted to life at the ocean's surface, but visitors frequent the area as well. Uranus may have had a hand in this because the skin of the sea is fertile ground, serving as an incubator for the eggs of fish and other marine species, and as a nursery for their young. For them the neustonic lifestyle is often just a phase and the offspring usually descend to the deeper levels once mature. Beforehand, however, they provide surface predators like *Halobates* with an important food source and influence the seasonal composition of the extremely dynamic and diverse SML. But there are unwelcome visitors to the SML as well – and they are bound to stay. The sea's surface is a dump for human-made pollutants in extremely high concentrations, from oil spills to heavy metals, from pesticides to microplastics. Many of these molecules and particles float in water and accumulate in the SML, where they encounter the eggs and young of so many species whose development they could disrupt.

The sea-surface microlayer is its own ecosystem, but it is still connected to the underlying waters, their inhabitants – and rising slimes. While microbes in the SML actively produce biofilm-like gels, at the base of the habitat is organic matter floating up, including gel-like particles. Their quantities and composition depend on the source, which in many cases is marine life on the ocean floor below. Corals, for example, produce enormous amounts of slime – up to five litres per square metre of reef. Some of it will rise up to the surface, and denser coral populations on the sea floor usually make for a more gel-enriched SML above.

But this is just another example of the influential afterlife slimy waste products can have in the ocean. Primarily, corals need their multi-functioning mucus for protective and hygienic purposes, like the casting-off of parasites and clogging particles, as Christian Wild of the University of Bremen discovered while working on his thesis in Australia. 'I sprinkled the corals with sediment only to observe how they then produced terrific quantities of slime,' says Wild. Ultimately, he was able to show that coral slime also supplies their diverse ecological communities with energy and nutrients. 'Altogether, slime produced by corals is a dominant component of the organic matter in the water column above reefs,' Wild explains. The slime acts as a sticky trap, catching planktonic organisms and other particles. The individual units grow heavier, eventually sinking, and can be broken down by microorganisms, which may be what keeps nitrogen and phosphorous in the very nutrient-poor system of coral reefs and makes them available again to a range of organisms.

'When I published these findings, I was initially mocked by my colleagues, who called me "Mr Slime",' says Wild. 'Now they're consulting me more frequently because it's becoming increasingly clear how important slimes are in the ocean.' And on the surface they act – with the help of their microbial colonizers – not unlike the mucus layers of our gut: selective exchange occurs at the interface, molecules diffuse and matter is broken down in a process of microbial degradation. It is a delicate balance that might fundamentally shift as a consequence of the climate crisis. Will the resident microbes in a warming ocean become all the more active, prompting the microlayer to swell? If so, would a thickened slime skin hamper or maybe even offer less resistance to exchange?

Even slight disturbances or small-scale changes could have devastating effects, given that they occur on an interface covering two-thirds of the planet's surface. Big problems demand big solutions that often pose a risk themselves. Some large-scale attempts, for example, aim to mechanically skim bigger pieces of plastic off the sea surface but could catch neuston as well. After all, both floating plastic waste and inhabitants of the water's surface are very similar in their behaviour and exposed to the same currents: 'Plastics mimic the neuston world,' writes the American jelly biologist Rebecca Helm in an article in *The Atlantic*. '[They're] buoyant, surface-bound and rubbery.'

Cleaning up 90 per cent of the plastic using the current skimming method means potentially destroying 90 per cent of the neuston, Helm writes. In the worst-case scenario, this could doom parts of the existing ecosystems and their adaptations to an anthropogenic world: *Halobates* sea skaters will just as happily lay their eggs on a floating plastic bottle as on hard-to-come-by driftwood or other flotsam. A single piece of waste was found bearing more than 70,000 eggs laid by this insect, one of the few known beneficiaries of the tide of plastic, which can now probably find suitable breeding sites all the faster thanks to the floating rubbish.

24

The Contours of the World

> *'. . . makes the wheel spokes fly, boys, when you fasten to him that way; and there's danger of being pitched out too, when you strike a hill . . . Hurrah! this is the way a fellow feels when he's going to Davy Jones – all a rush down an endless inclined plane! Hurrah! this whale carries the everlasting mail!'*
>
> Herman Melville, *Moby Dick*

'Do you fear death?' Davy Jones asks the sailor, who quakes with fear. 'Do you fear that dark abyss?' With his betentacled face, the heartless baddie from Hollywood's second *Pirates of the Caribbean* film might look like a younger brother to Lovecraft's evil Cthulhu, but he has his own, much more ancient history. Is he the devil himself or just the captain of the ghost ship *The Flying Dutchman*, cursed to traverse the seas for evermore? The myth's beginnings are murky, but the Davy Jones figure was feared for centuries; all sailors who drowned or died by other means on the open seas belonged to him. As a literary figure, he proved irresistible to many writers, from Daniel Defoe to Herman Melville, who has the sailors in *Moby Dick* trembling at the thought of a watery grave in Davy Jones's domain.

As the modern Hollywood reincarnation, he offers a treacherous choice to his captives, who are close to death or have already crossed the threshold: instead of floating forever in the depths of the ocean, they can be taken aboard his ship as slaves for the next 100 years. The captain is vulnerable himself because he has cut his own heart out of his body and keeps it in an elaborately locked trunk. In the original myth, however, Davy Jones's locker has a darker significance as a metaphor for his kingdom, the sea floor, a watery grave for dead seafarers and a graveyard of sunken ships.

As Rachel Carson writes, there was yet another version of the story: the denser seawater in the ocean's depths and the buoyancy of sinking bodies and wrecks were said to slow them down, bringing them to a standstill some distance above the sea floor. If this were true, dead sailors and sunken ships would float forever, ghostlike, deep beneath the waves, neither finding rest on the sea floor nor coming up to the surface again. The physical considerations for this fictitious phenomenon don't really work. However, there are actual differences in salinity and density between layers of the oceans and also compared to other bodies of water which may determine the limits of some habitats. It is but one example of different spheres meeting – and shaping the contours of the world.

An estuary is where the mouth of a river meets the sea, with a tongue of saltwater pushing upstream at high tide like a wedge, without fully mixing with the freshwater. The denser seawater usually remains at the bottom, bringing nutrients and oxygen, with the lighter freshwater rising above and expelling waste into the sea. Depending on the type of estuary, there can be a rather abrupt salt front working its way upstream. How do organisms in these open, yet separate, spheres perceive one another? Perhaps they float past each other, eyes wide, much like human visitors wandering through tunnels of glass in modern aquaria, staying in their own medium while perceiving the underwater world from up close.

Fresh- and saltwater meet at an interface that defines one of the most productive and diverse but also vulnerable ecosystems, though with a host of different habitats. Some inhabitants only tolerate areas with narrowly defined conditions such as a specific salt content. Among them are marine species that bury themselves in the sediment when the tide ebbs, to escape the approaching freshwater.

And from that separation new habitats form: adaptations like these allow marine specialists, just like less sensitive species, to advance far into the watercourse, as the ichnologist Anthony Martin writes. Estuaries are a class of ecosystems of their own.

Seen in context, estuaries are one of the ocean's marginal or peripheral biotopes. It was for these areas that Yuvenaly P. Zaitsev, studying the neuston, coined the now seldom-used term 'contour'. This refers to the edges of the oceans, such as the seabed, the surface and the coastlines, which are repeatedly flooded with saltwater in the shifting of the tides. The species inhabiting these borderlands are what Zaitsev called 'contourobionts', forming, despite their differences, an ecological group, connected by their equally challenging lives at marine interfaces.

Zaitsev based his ideas on the work of the Russian scientist Vladimir Vernadsky (1863–1945), who has quite unjustly been forgotten in the West, even though he deserves a place next to better-known scientists. 'Indeed, Vernadsky did for space what Darwin had done for time: as Darwin showed all life descended from a remote ancestor, so Vernadsky showed all life inhabited a materially unified space, the biosphere,' wrote Lynn Margulis and Dorion Sagan in their book *What is Life?* He was certainly ahead of his times, as a scientist – and as an idealist.

In *Stalin and the Scientist*, Simon Ings writes that Vernadsky, coming from a wealthy family, decided along with several friends to devote his life to the good of the Russian people. The friends swore an oath to produce as much as possible, consume as little as possible and to treat the needs of others as they would their own. Their idealism proved itself when drought, flooding and other climatic disturbances left 35 million people facing famine in 1891. These extreme conditions affected the fertile regions known for their humus-rich black earth, which the wind swept up in such great quantities that it blocked out the sun.

While Vernadsky, his friends and other members of the aristocracy were setting up makeshift emergency kitchens and working to ease the country's suffering, the government was selling what little grain remained to struggling states in Eastern Europe. The British historian Orlando Figes sees this hunger crisis, the failure of state power and the strengthening of civil society as a crucial turning point. This public awakening is one of the broad social changes which laid the foundations for the coming revolution.

Vernadsky is now mainly considered as one of the fathers of biogeochemistry, in part thanks to his groundbreaking contribution to science. A mineralogist by training, his specialist area was the geology of our planet, which he recognized – based on observations made while on a hike in the Alps – as the close link between the history of the cosmos and life itself. How do life and inanimate matter influence each other and how are they connected? More than 100 years before Lovelock's Gaia, Vernadsky had already developed a large-scale and rather modern concept of the biosphere, a term he didn't coin but picked up from a colleague.

He described oceans as concentrations of living matter which forms a kind of film that covers land and marine interfaces, among others. This is not what he meant, but we know now that biofilms and other slimes produced by that living matter shape and stabilize our planet's contours. These are the soft edges of the sea, and we need to understand how fine material like sand particles and mud are moved by the water currents if we want to assess how the climate crisis will affect estuaries, the coastal zone and the continental shelves, the underwater bases of the continents – which make up 8 per cent of the world's oceans.

These contours are critical for human activities that depend on open channels for ships and seabeds suitable for engineering projects like the laying of undersea cables – a challenge since the 1850s. Yet the underwater profiles of rivers and the seabed are not static. The soft sediment can form dunes, arrange itself in gentle ripples or lie flat and level. Depending on the profile of the dynamic beds, the water's movement, speed and direction can change and will do so even more in the future. 'These coastal and nearshore locations are among the most sensitive regions in terms of sea level rise, a problem exacerbated by the predicted increased frequency of extreme weather events, which will act to alter a range of sediment transport processes,' Daniel R. Parsons, of Hull University, and colleagues state in a publication.

But which factors determine subaqueous bedforms? For quite some time, computer simulations were of little help, calculating more pronounced profiles than those which scientists observed in reality. These models were based primarily on sediments composed of sand, with grains that move freely. As it turned out, some key ingredients were missing from the calculations: sediments are actually made up of sand, weakly cohesive mud and biological life forms that produce a lot of very sticky mucus. First tests confirmed the goo's influence on bed forms: sand only tends to produce ripples and dunes, which were reduced when clay was factored in. But the beds remained virtually flat when enough sticky slime was added.

Biological goo is capable of cementing particles together in the beds of streams as well as in coastal regions, stabilizing the sediment. And a little slime can go a long way, even flattening out underwater beds. This speaks of an astonishing degree of adhesive power. But where exactly do the gels come from? It's the usual suspects who do the bulk of the work here. It's mainly microbes like bacteria and diatoms, which secrete exopolymers to cover the sediment with their sticky biofilms. But there are also worms and other burrowing invertebrates which tunnel through the sediment that makes up their habitats, peppering it with slime as they go.

The edges of the soft seabed are so important because patterns like dunes or ripples influence the direction and speed of the current in these areas differently than level sediments. In estuaries, for example, navigability of the waterway depends on the profile. People use these coastal regions in very different ways, so they are obliged to keep an eye on the sedimentary profile and be able to predict how it will develop in the future. This is of great importance because climate change is expected to cause sea levels to rise further, which could well upset the delicately balanced processes at these interfaces.

But it's mainly the microbial masses that keep the sediment together – and bring unexpected visitors to the table. At low tide, mudflats may be exposed that, due to their high proportion of organic matter, squelch with every step visitors take. It's a desolate landscape at first sight, yet Canadian scientists observed thousands of migrating sandpipers stopping over. What were these birds eating? Examinations of their stomach contents brought up nothing but water and sand. Their brush-like tongues later solved that puzzle: sandpipers, and perhaps other migratory birds, harvest the energy-rich slime that covers the flats.

It's mainly the extensive biofilms that diatoms produce here that keep the microbial neighbourhood together, offer food to invertebrates, crabs and fish but also keep these migratory birds on endurance flights aloft. This could be due to essential fatty acids in the goo, whose composition varies throughout the year. It seems that the birds depend on these crucial lipids, and might even seek out specific stops at times when microbes there have that source of energy on offer – a kind of 'lipid pathway' for the birds to follow. Without these single-celled algae, the whole system may collapse before we even understand what makes specific mudflats valuable resources for wildlife.

How do these microbes react to changes in their environment? Inhabitants of estuarine sediments have to adapt to a world that changes cyclically between fresh- and saltwater. It's the salinity that counts in biofilms since it has a marked effect on the adhesive power of gels. Experiments have shown that biofilms also stabilize sediments in freshwater. However, they are probably less sticky than marine slimes. At least, this is true until the seawater rushes into the estuary; could the higher salinity affect the adhesive power of existing biofilms? Or will marine microbes, with their superior gelling capability, simply take over? We don't really know yet.

Seawater spreads its microbes far and wide in coastal meadows and salt marshes as well. These ecosystems are found further inland, but they too are periodically flooded with saltwater – and being shaped and glued in the process. I visit one such habitat on Cape Cod, where swaying boardwalks snake their way through the tall grass. I'm able to walk down to the beach without getting my feet wet, but on my return the boards have already been flooded. The saltwater always brings in sediment and any creatures which inhabit it. If they can tolerate this fluctuating habitat, washed-up microorganisms will settle here and produce slimes which make a significant contribution to stabilizing these sensitive zones.

It is essential that such wetland habitats are preserved, not only for their unique nature but because they act as buffer zones between land and sea, protecting the coast from erosion while also filtering out fertilizer run-off from the fields before it reaches the sea. Many of the contours of the planet are hybrid zones between river and sea or ocean and land; they keep a fragile balance that the climate crisis has already begun to disrupt by pushing the rising, warming and acidifying sea further inland and upstream. This is a process that could change the face of the world when salt marshes find themselves closer to the ocean and reduced to tidal zones resembling mudflats. Similarly, if rivers gradually run dry, it will upset the balance in estuaries. Anthony Martin, an expert in trace fossils, writes of barnacles and other marine species found living thirty kilometres inland in Georgia, US. They settle in a body of water, which remains a river in name only.

But there's another marine interface to consider too. This one exists vertically, where the deep sea comes into contact with goo from the upper layers of the ocean. Just as the edges of the sea are crowded with life, so the light-filled upper layers teem with photosynthesizing phytoplankton. These organisms are busy taking up CO_2 from the atmosphere to produce – with the help of sunlight – organic matter and oxygen. Diatoms alone release at least 20 per cent of all the oxygen we breathe. They might not stay the night, but never mind, enough plankton remains to warrant the biggest migration on Earth. As soon as the sun sets, zooplankton, molluscs, fish, crustaceans and other animals from the mid-ocean twilight zone rise to the surface layers of the sea. Here, under cover of night, they feast on phytoplankton, only to disappear to the depths again at daybreak.

This seems like a rather straightforward food chain, but in fact it constitutes just one part of the marine carbon cycle which is an essential climate factor. Phytoplankton photosynthesizing means that enormous amounts of CO_2 – an anthropogenic greenhouse gas – gets removed from the atmosphere and broken down. Its organic carbon ends up in biomass such as ever-bigger predatory bodies or faeces and other waste. And a lot of that diverse carbon-rich material sinks down to the sea floor, where it may remain for thousands of years or even longer. In other words: the more biological garbage finds its way into the deep, the greater the amount of organic carbon that won't end up in the atmosphere as greenhouse gas, at least not so soon.

We know now that gelatinous zooplankton – jellyfish, ctenophores and pelagic tunicates like salps – play an outsize role when it comes to this biological pump. Their bodies soak up so much organic carbon that it has been termed 'jelly-C'. When they die, their bodies often sink so fast that they can't be degraded by microbes in the upper layers. What's more, jellyfish alone can create gigantic blooms that cover enormous areas and end in C-rich jelly-falls. And the rest? Organic matter in the sea always includes mucus, biofilms, gel-like particles – in short, slimy stuff that sticks.

Inadvertently, this acts like a glue trap, fishing other particles from the water till 'flakes' emerge that are big and dense enough to sink as 'marine snow'. Marine snow is a rich organic rain falling through the sea. Part of it gets eaten on the way down, or microbes might add their slime to the particles, which become little ecosystems with different microhabitats. What escapes hungry mouths in the deep will add to metres-high deposits of sediment. But only as long as the biological pump keeps running at a steady rate and is not being disrupted by the climate crisis.

There is even a theory that Huxley and Haeckel's ur-slime sample might have contained some marine snow from the deep which would have actually warranted their attention. Perhaps the slimy flakes of gently sinking marine snow are the true riches contained within Davy Jones's locker: 'When I think of the floor of the deep sea, the single, overwhelming fact that possesses my imagination is the accumulation of sediments,' Rachel Carson writes in *The Sea Around Us*, continuing: 'I see always the steady, unremitting, downward drift of materials from above, flake upon flake, layer upon layer – a drift that has continued for hundreds of millions of years, that will go on as long as there are seas and continents. For the sediments are the materials of the most stupendous "snowfall" the Earth has ever seen.'

25

The Earth's Living Crust

In the morning the dust hung like fog, and the sun was as red as ripe new blood. All day the dust sifted down from the sky, and the next day it sifted down. An even blanket covered the earth. It settled on the corn, piled up on the tops of the fence posts, piled up on the wires; it settled on roofs, blanketed the weeds and trees.

John Steinbeck, *The Grapes of Wrath*

A haboob is a dust storm. It's an Arabic word for a phenomenon that struck the American Midwest like a plague of biblical proportions just under a century ago. This storm was not the work of God's chastising hand, though; it was the worst man-made environmental catastrophe the US has ever seen. East of the Rocky Mountains, the Great Plains extend like wide corridors all the way from Mexico to Canada. The land is dry and barren, naturally covered by a community of robust plant life – primarily prairie grasses – whose dense root networks stabilize the soil beneath. But then the humans came. The government knew the plains were not arable land and that, at best, only large farms with complex irrigation systems might be able to make a living there.

Little family farms without adequate capital, however, would not be able to survive. Nevertheless, the land was advertised intensively into the 1930s and allocated to settlers free of charge as fertile land requiring little more than a 'tickle' with the plough to unleash its productivity. The railways, needing to pay off their newly laid network of tracks, joined in the propaganda for the 'Nile of the New World'. In the decades to come, around a third of the Great Plains would be transformed into green grassland and pastures, worked intensively with horse and plough, which disturbed the dense root network of the original plant life. As we now know and will discuss later, essential biological glues were lost as well.

Another of the government's promises was that the rains would follow the plough, but what actually followed was, in essence, the apocalypse. The exposed earth grew drier and hotter, losing its stability, and became subject to droughts and erosion. This was followed by a rare weather phenomenon, reversing the jet stream which usually carried clouds and rain towards the Midwest. Now the rains stayed away, the harvests withered and vast clouds of dust loomed like a black rock face, kilometres high. In 1933 alone there were more than fifty haboobs, which swept across the landscape.

Images from this period show houses and farms submerged as if beneath a grey flood, waves of dust and earth washing as high as their roofs. Contemporary witnesses spoke of fine grains grating against the skin like sandpaper, blinding people, suffocating cattle and leaving children sick with the 'brown plague', a type of pneumonia caused by dust, as described by Timothy Egan in his award-winning book *The Worst Hard Time*. There was no escape. In the evenings, families sealed their windows and doors with damp towels, yet they would still find themselves shovelling dust out of their homes and cottages come morning.

The dust in the air muffled the voices of crowing cockerels and the sun hung blood-red in the sky. By afternoon it would be dark again, and anyone out and about in the dense haze would tie a rope around their middle so as to be able to find their way back. The dust was carried from the Great Plains as far as Chicago at the other end of the country, and all the way to the capital in Washington, such that it left a brown coating even on ships at sea. The 14th of April 1935, 'Black Sunday', brought the mother of all haboobs, during which, according to Egan, twice the amount of dust went swirling across the country in a single afternoon as had been dug up over seven years to build the Panama Canal.

For most farmers, there was no longer a possibility of making a living in the Midwest. The photographer Dorothea Lange became famous for her portraits of careworn and gaunt migrants and their rag-clad children, making their way westwards. Like the Joad family in Steinbeck's masterpiece *The Grapes of Wrath*, they were following another empty promise, that there would be work in the big cities of California and elsewhere, a promise thwarted by the global economic crisis which shook the 1930s. Those who stayed behind had little more luck, as the meagre harvests were destroyed by devastating blights and, in any case, the land did not recover from the destruction it had suffered.

The consequences are not always so catastrophic when dry soil loses its stability and is exposed to erosion. But even little changes can jeopardize our food supply if they occur on a sufficiently vast scale, if, for example, entire areas of land are exposed to higher temperatures and lower levels of precipitation due to climate change. Most at risk here are the biological soil crusts, ecological communities often unseen or mere millimetres tall which cover the ground in deserts and dry regions, but are also capable of growing on and underneath stones.

Where and whether they form depends on precipitation, temperature and the agricultural use of the land. They are particularly prevalent in deserts, as well as steppes and savannahs, especially in southern Africa, Australia and Asia, and in the Southwest of the US. They are seldom found in temperate zones, such as those in Central Europe where vascular plants like shrubs and trees completely cover the ground. According to a study by the Max Planck Institute for Chemistry in Mainz, led by Bettina Weber, biological soil crusts cover up to 12 per cent of the Earth's surface, corresponding to 40 per cent of actual land mass.

If layers of gel form a kind of mucosa on the surface of the world's oceans, then biological soil crusts are the Earth's living skin, a protective barrier. Yet they also accumulate and transform nutrients and play a key role in larger biogeochemical processes, such as the global nitrogen and carbon cycles. 'The Earth's crusts are dry, hard and, well, crusty,' says Weber, 'but there's also a certain sliminess.' Cyanobacteria are pioneers of soil crusts, setting things in motion by secreting exopolymers which will, as usual, build a slimy matrix. It's sticky enough to glue particles in the soil together, protecting it from erosion. This hydrogel also binds and stores the small amount of water that is present – for example, after rain – before it evaporates or trickles away.

Biological soil crusts are complex ecological communities which science categorizes according to their developmental stages. Cyanobacteria are the vanguard, paving the way for other microbes like bacteria, as well as archaea or fungi, which join the young community, doing their part by breaking down organic matter. These might be followed by lichen and mosses, possibly even by worms, slugs, snails or springtails and other arthropods as well. It can take years or even decades for a soil crust to become this diverse, potentially boasting many hundreds of different species.

However, there are differences between these communities in the crust, not only in regard to the level of maturity they achieve over time, but spatially too: soil crusts are separated into layers. Strongly pigmented fungi and all photosynthesizing organisms, such as cyanobacteria, generally colonize the top layer because they require and can tolerate UV radiation, while shadier characters live lower down in the ecosystem. They must all be able to survive long periods of drought. Specialists in this area include, for example, the tardigrade, which can enter a state of dormancy before quickly reactivating as soon as water is available.

These days, however, biological soil crusts are under threat. According to Bettina Weber's calculations, up to a quarter of this protective coating could soon disappear. Climate change is as much a factor as population growth, which requires the expansion of arable land to include dry and previously unused strips, hitherto covered by biological soil crusts. This development could have consequences across the globe, affecting the nitrogen cycle, among other things.

Nitrogen is present in the ground and the atmosphere but cannot be taken up directly by plants.Plants rely on soil microbes which fix nitrogen, making it available for plants to use. Bettina Weber has calculated that as much as half of this essential service may be provided by soil crusts. Disturbances to the structure would hit many ecosystems in nutrient-poor regions particularly hard. But the loss of the soil crusts would also expose the ground to intensified erosion by water and wind, enriching the atmosphere with the tiniest of particles. It doesn't need another dust bowl like that of the Great Plains in the 1930s to pose a risk to human health, and not just for people with allergies and hay fever.

The notion of miasmas, transmitting fatal infections such as malaria (from *mal aria*, bad air), has now been put to bed.Yet the air around us is filled with microbes, pollen and other particles which have the potential to cause us harm.The great microbiologist Louis Pasteur was the first to prove that open wounds could be infected with germs from the air. In a sense, this made him the founder of aerobiology, a discipline which witnessed its first and – to date – last golden age in the 1930s, when farmers in the Midwest were facing a global financial crisis, devastating haboobs and plant pathogens thrown in for good measure.

Fred C. Meier of the US Department of Agriculture happened to be the right person in the right place at the right time. A tremendously charismatic man with a pilot's licence, he hoped to discover how the deadly rust fungus – or its spores – was spreading, and to what extent weather and the atmosphere were contributing factors. To this end he recruited American aviation's shining stars, including Amelia Earhart, who later died in an air crash, like Meier himself. She was joined in her aerobiological efforts by a celebrity couple, the Lindberghs. Charles Lindbergh's pioneering flight across the Atlantic overshadowed his wife's success somewhat, though

Anne Lindbergh was one of the first female pilots in the US and steered the plane on their joint flights as well.

And it was a good thing she did. In 1933 the couple flew from the US, over Greenland and as far as Denmark – while being on a mission. As discussed with Meier beforehand, they used 'sky hooks' as airborne traps. Charles had constructed them himself out of a metal cylinder containing oily, sticky glass slides which would catch solid particles in the air. In fact, a kilometre above Greenland they found spores of exactly the same rust and other fungi which were growing on the ground thousands of miles away, causing vast agricultural damage. The findings were clear: these spores were nomads which travelled by air, high up in the planet's atmosphere. And they were not alone: the Lindberghs also collected grains of pollen, fragments of fungal mycelium, single-celled algae like diatoms, insect wings, volcanic ash and glass particles in their sealed traps.

Like the microbes of the deep biosphere in the Earth's crust, other bacteria and spores define the limits of life high beyond the clouds. The living inhabitants of the air, which drift with the wind and cannot fly themselves, are sometimes referred to as aeroplankton, inspired by the ecological communities which float through the oceans. This is where the similarity ends, however. The lofty heights are – unlike the sea – not home to a hidden and crowded world of gelatinous creatures. There are no see-through fish of the air or transparent air-dwelling jellyfish, even though some creative minds wished there to be.

'Gilliatt did not believe that the air was an uninhabited desert,' Victor Hugo writes in his *The Toilers of the Sea*:

> Since the sea is full, he used to say, why should the atmosphere be empty? Air-coloured creatures would disappear in daylight and be invisible to us . . . For since the light would pass through them, giving them no shadow and no visible form, they would remain unknown to us and we should know nothing about them.

Gilliatt imagined that if we could drain the earth of atmosphere, and then fished in the air as we fish in a pond, we should find a multitude of strange creatures.

The air is home, however, to numerous microbes, and we already know, to some extent, where they come from, or at least where their journey begins. They can find their way from the ocean into the atmosphere when air bubbles rise through the water and burst at its gel-like surface, which is densely populated by microbes. Astoundingly, a similar phenomenon can be observed when raindrops strike farmland, the impact causing dust and microbes to swirl up into the air.

Even the leaves of plants can be a starting point for propelling matter into the air, particularly when the plants are under the weather. Many pathogens which affect humans are transmitted via the air we breathe, or via coughs and sneezes – as the coronavirus pandemic has taught us all too well. This is known as droplet transmission. Lydia Bourouiba at MIT demonstrated that plants spread pathogens in a similar way, with fungi, for example, travelling via spattering droplets of rain. They cover themselves in a slimy coat of mucilage, which protects them and prevents them from being carried high up into the air on the wind. If a raindrop strikes an affected leaf, the water splashes off, carrying the pathogen with it, maybe to its next host.

Pseudomonas syringae is an economically important pathogen infecting hundreds of plant species which also specializes in life in the air. This bacterium is present across the world, including in water, but can also survive for several days in the atmosphere, where it is thought to live on fragments of plant matter swept up into the air. We do not know if it forms a biofilm here. 'So far, we lack the necessary evidence,' says the microbiologist Hans-Curt Flemming. 'But it is very probable that these kinds of biofilms exist. Microbes could live in the clouds in little clumps.'

Or they fashion themselves fancy airships: Spanish researchers have shown that microbes can travel ensconced in atmospheric dust, even between continents. Their vehicles of choice are iberulites, dust particles made from different minerals that reach considerable size – and are glued together by bacterial slime. These kinds of aggregates from mineral and biological components occur all over the world. The ones that were studied in detail this time were found in the city of Granada but held dust grains and microbes from the Sahara. In the atmosphere they had been caught in a water droplet as a bioaerosol and had then taken on the characteristic shape of iberulites, a little like a dented cannonball. What primarily keeps them together is slime produced by the bacteria, which not only procure an airborne vehicle that way but also find nutrients inside the aggregate and are protected from UV-radiation and other dangers.

Some strains of *Pseudomonas syringae*, however, have the means to travel straight down to Earth. They produce a protein which causes water to freeze at unusually high temperatures. Like other microbes and particles, this bacterium acts as a crystallization point for ice formation. One hypothesis is that *Pseudomonas syringae* may be able to return to the ground inside a self-made hailstone or snowflake when conditions at altitude become too uncomfortable. Thanks to its freezing proteins, a harmless version of the pathogen is also being used to make artificial snow.

Other microbes could even be directly involved in fundamental processes in the atmosphere. There is already talk of an airborne microbiota that might change with the seasons. A long-term study has demonstrated that the microbial composition of aeroplankton over Greenland is highly dependent on the time of year, exhibiting vast differences in summer and in winter. Alongside spiders, wingless insects, spores, plant seeds and many other inhabitants and particles, it is home to hundreds of species of bacteria and thousands of species of fungi. Yet climate change threatens irreversibly to change these interactions and communities before we get a chance to decipher them and understand the role they play in the formation of rain, hail and snow in the hydrological cycle and other global processes.

It's not that hard to lob microbes into the air; a simple cough will do the trick just as easily as bubbles bursting on the ocean's surface. But soils – and especially their biological crusts – are closely tied to aeroplankton too. And they're equally threatened by disruption through climate change. The expansion of agriculture is another danger, and even smaller damage can have lasting effects. Shoes, hooves and tyres are capable of destroying these fragile biocrusts, which may take decades to regenerate if they ever get a chance. To lose them would mean losing what were probably evolution's first ecological communities. Not only do they occur on all continents and in all climatic zones, but they were probably the first to venture on to dry land, forming along the edges of bodies of water before moving further inland. They still play a vital role in shaping the habitat of many other organisms, by fixing nitrogen and binding CO_2 from the atmosphere. They are also important for weathering processes, breaking down mineral underlayers.

Since the cyanobacterias' slimy matrix stores the little water there is, the soil-crust community and higher plants benefit, even on grazing land. Sometimes, however, less is more. The Atacama Desert in Chile is one of the driest places on Earth. Very few bacteria, algae, fungi and lichen are able to survive here, in the soil crusts or as part of the soil microbiota. When the first rainstorms for decades made their way across this region in 2017 – a consequence of climate change – it seemed that the born survivors which inhabit this area would finally be granted a well-deserved embarrassment of liquid riches. In fact, the episode culminated in a microbial massacre as the unprecedented excess of water caused the organisms to burst. Of the microbes which normally occur in and on top of the soil in Atacama, only a handful of species survived.

26

Medea and the New Era

Some might find this new view of life depressing. I find it exhilarating, for if correct, only we humans (or other intelligent species out there in the cosmos) can change the rules and save the rest of life, as well as our own species, from itself.

Peter Ward, *The Medea Hypothesis: Is Life on Earth Ultimately Self-Destructive?*

'The killer is life itself,' writes the palaeontologist Peter Ward. 'It will hasten the ultimate death of all life on Earth.' Ward named this apocalyptic vision Medea after the sorceress of Greek myth who murders her own children. He also named it Medea in deliberate contrast to the Earth mother, Gaia, namesake of James Lovelock's theory of life supporting its own continuation and survival. Ward considers this theory as doomed as the notion of ceding control to Mother Nature – and sees multicellular life on a suicide mission. In his view, we fail time and again to keep microbes on a short lead. Left unchecked, they spread rapidly and get a chance to set processes in motion which could bring an end to their only competition: us. Ward traces mass extinctions in the Earth's history back to the 'nefarious poisoning of land and sea' by microbial processes.

But we mustn't fool ourselves; humans don't come off much better than microbes, as we are in danger of reducing the lifespan of the Earth as an inhabitable planet. But not all is lost, according to Ward; we have it in our power to take the reins and intervene in global cycles. Thanks to our intelligence and ingenuity, we might just be able to defeat Medea. This dystopian yet hopeful vision is somewhat reminiscent of one of Vladimir Vernadsky's propositions: the idea that the influence of human thought could be as powerful as life and its biosphere. Among the achievements of this proposed noosphere are language, culture and technology. Now, the internet probably represents its most profound and globe-spanning expression. Unfortunately, we rarely think more than a few generations ahead, a myopic view which has and will cost us dearly in terms of climate change, environmental destruction and species extinction.

There is still some debate as to whether we should officially declare this current period the era of the Anthropocene, the geological epoch in which human influence shaped the planet itself. And if we do, should we let the Anthropocene begin with the First Agricultural Revolution some 12,000 years ago or in the 1960s? How we define it matters little in regard to the profound disturbances and disruptions we are facing now. Future generations will have to deal with a legacy that includes deposits of dangerous nuclear waste, but also plastic particles that are being found in every nook and cranny of the planet, from the Mariana Trench to Mount Everest.

We live in a rapidly changing environment which will affect every living being, including us humans. What will the future hold – and what role will slimes play in it? They could help save us from our own environmental follies, or stifle whole ecosystems. They could give us inspirations for novel eco-friendly applications, or be lost before we even know about them. From technology to biomedicine, they offer a flexible Lego system of seemingly base hydrogels with an endless supply of extra pieces to customize them for every need and preference. Our own mucus protects us, but it can contribute to an unhealthy imbalance in our bodies. In short, slimes can offer new ways to connect us with ourselves, other life and the environment – or make already bad situations catastrophically worse.

Nuclear waste, for example, needs to be contained in an inert environment for as long as it takes to lose its destructive radioactivity. But microbes will always settle there and interact with their surroundings. After all, they have a penchant for extreme environments, such as the former underground uranium mine Königstein near Dresden, Germany. Heavy metals, radioactive particles and acid waters: what's not to like? Scientists found 'snottites' there, stringy biofilms that look like slimy stalactites. They protect microbes and allow them to be active, which could have unforeseen consequences if they interact with the dangerous waste. Unless, of course, we use their seemingly detrimental properties to defuse and transform our most toxic legacies.

When it comes to microplastic particles, the extraordinary power of slime can be a liability. Designed to be sticky or to filter organic matter, all kinds of biological glue traps might hold on to microplastics as well. We know from studies on periwinkles that gastropod pedal mucus can retain microplastics. If other snails forage on the slime, they take up these particles as well. They release a substantial amount of them in their faecal pellets, which might prevent them from adverse effects, but could pass these excreted microplastics on to whatever other foragers in the food web ingest them.

In the ocean, however, many species employ high-functioning mucous traps to capture food suspended in the water. And their sophisticated designs could just as well accumulate and transport microplastics. These particles have already been found in nearly all types of marine environments and in a variety of species across different food webs. And some filter-feeding animals have already been proved to ingest microplastics, giant larvaceans (*Bathochordaeus stygius*) being but one example. These tadpole-like animals live inside extensive mucus houses that filter microplastics of different sizes which can then be taken up by the animal – and passed on. Many mid-water and deep-sea animals eat larvaceans, their faecal pellets and cast-off houses, taking up bound microplastics at the same time.

Or might there be a 'gelatinous solution to microplastic pollution'? GoJelly is a project funded by the European Union that aims to use jellyfish as human food, fish feed and fertilizer, with a particularly prestigious job reserved for their mucus as a biofilter for microplastic particles. Jellyfish produce slime in copious amounts when stressed or hungry. *Cassiopea xamachana* lies 'upside-down' on its bell on the sea floor and lobs mucus bombs in the water that will sting unaware swimmers and even bring down prey. GoJelly's mucous filter, on the other hand, would catch microplastics in wastewater plants, for example, to prevent them from ever reaching the sea.

But why stop with jellyfish? Hydrogels offer solutions to all kinds of complex problems thanks to their ability to give in. Traditional robots, for example, are rigid, which makes them somewhat inflexible, possibly even dangerous around humans. Soft robots, machines and devices, are different, they move freely and are safe to interact with. Some of these soft newcomers are already picking up speed, often with a particularly delicate touch. There is one gel-based robot, for example, that works well underwater and can grip fish or even more delicate animals without hurting them. Another robot can move on land, pick up cargo and even climb hills. All these and more examples are based on hydrogels. And soft robots might actually be closer to the original meaning of the word. Karel Čapek not only wrote about newts at war, he also wrote the play *R.U.R.*, in which he used the newly coined term that his brother had come up with: 'roboti' were artificial humans, similar to biological flesh and blood, created synthetically from slimy protoplasm.

And this is only the beginning; due to their compliant and flexible nature, tailor-made hydrogels could offer an astounding variety of properties: they could be transparent, stretchable, locomotive, biocompatible, remote-controlled, weavable, wearable, self-healing and shape-morphing, 3D-printed or improved by different ingredients. Their already smart functions might be manipulated on demand, as hydrogels respond to stimuli like light, temperature, pH, magnetic fields and electric currents. Most of these multifaceted hydrogels are still far from actual application. But the outlook is not bad for a simple water-swollen material.

When it comes to interfaces between humans and hydrogel, biomedical research is especially interested in wound dressings. How do you replace damaged skin if a graft is not an option? Hydrogel-based substitutes can mimic human skin and be applied in different ways – or even sprayed on. As for possible upgrades, smart hydrogels with tailor-made functions could act as platforms for cells, antimicrobial agents, growth factors or other molecules. Integrated sensors could then release drugs and monitor progress such as inflammation levels. Whether they are applied permanently as skin substitutes or only temporarily as dressings, smart hydrogels are expected to deliver a new approach to wound care. They stand to revolutionize other biomedical fields as well. For example, contraception: in the future, hydrogels could be used to block the passage of sperm in the male body. It would be a reversible solution, because the plug could be dissolved whenever necessary.

Another focus of medical innovation is the restitution of dysfunctional mucus barriers in our body and – if necessary – their microbial colonizers. Since there is no ideal microbiota for any one person, it is hard to define the therapeutic application. But we know that our internal slimescapes are shaped by eager gardeners like *Akkermansia muciniphila* in the gut, which as a species makes an attractive target for probiotic approaches. In the female genital tract, *Lactobacillus crispatus* could be a cornerstone species that makes the environment unpleasantly acidic for pathogens and strengthens the slime barrier. Otherwise, pathogens will seize their chance, breaking down mucins and weakening the hydrogel. That kind of dysbiosis can result in so-called bacterial vaginosis, which is associated with a higher risk for sexually submitted diseases, infertility and pre-term births. One study suggests that an invading microbiota could be the cause. It's found in the women's male sex partner – in the penis.

Overall, it's hard to say what causes microbial dysbiosis in the human body. It seems that the environment frequently has a part to play, leaving our mucus barriers and their microbial inhabitants under attack from many sides at once. Our diet is one crucial factor since it provides food for our microbiota as well. The typical Western-style indulgences are often overly rich and lacking in fibre. What are microbes to do when on the lookout for complex carbohydrates? They will resort to breaking down our sugary mucus. But there is another possible risk factor. Studies on mice have shown that a common colouring agent used in foods is damaging to microbes in the gut. Titanium dioxide, or E171, is used as a whitener in many products, from chewing gum to toothpaste to cheese and sweets. This substance is not inert in our digestive tract; it changes the behaviours of our bacteria. E171 is likely to be just one of many disruptive elements – including microplastics.

We know that we ingest microplastics since they have been found in human excretions. Their possible effects remain to be seen, but some scientists fear an industrialized microbiota in our guts. What they mean is that our microbes could readily adapt to new environmental influences, even pollutants, pesticides and processed food, while we stay behind because genetic change is much slower. We are holobionts and have co-evolved with our microbiota. Will we still fit together when our own biology can't keep pace with them? According to the theory of industrialized microbiota, a growing gap between our slow genetic and the much faster microbial change could contribute to modern diseases like chronic inflammation, diabetes and cancer.

To return to the image of the hedgerow once more; the population of our mucous thicket changes dramatically as the microbiota is impoverished and its balance disrupted, in part because microplastics, pesticides and other 'unnatural' particles end up here. Are our slimescapes becoming 'edgelands' of sorts? These are fringe areas, the frayed edges of cities littered with decommissioned industrial plants and abandoned houses. But nature is not to be defeated, as a green wave surges from the surrounding countryside and reclaims these developed areas. Unique and in their own way complex ecosystems then begin to take shape in the environment, and they might well colonize our guts in the future too.

But it's not only human hydrogels that are in danger of changing; others might vanish for good. With every species that becomes extinct, a variety of presumably tailor-made slimes are lost as well – and with them a chance for novel applications. Antibiotics are highly efficient drugs that lose their edge because of overuse and the rise of resistance in pathogens. All higher organisms fight microbes one way or the other, amphibians among them: as a full-body exercise. Many species coat their porous skin with slime that is laden with pathogen-fighting substances. Might some of these antimicrobials be applicable for us as well? Possibly, but we might never know in many cases since amphibians are devastated by fungal diseases that might have severely decimated hundreds of species or even wiped them out.

This puts extra pressure on that particularly fascinating group of vertebrates, but it might only accelerate what climate change and other environmental crises like habitat loss would have achieved sooner or later anyway. The loss of biodiversity is a loss of precious resources, even if we don't know about it. Conservation efforts massively favour charismatic and supposedly higher fauna. A report on wildlife conservation by the European Union, for example, showed that investment per vertebrate species has been 468 times higher than that for invertebrates. When it comes to the preservation of biodiversity we're playing favourites – who happen to be birds or mammals in most cases. Who loses out? All the slimers, creepers and crawlers.

Amphibians get some of the limelight, at least compared to Mother Nature's most unloved children – the parasites. They are an integral part of all ecosystems: it is estimated that at least 40 per cent of all animal species, possibly many more, live in or on but always off other organisms. They influence or create the majority of links in food webs, even if we only know a minority of parasitic species and their often extremely complex life cycles. Scientists have recently presented a parasite conservation plan and proposed global co-operation in that regard. They will have to hurry, though: parasites are particularly vulnerable in their dependence on other species that could die out or shift their schedules or habits as a consequence of climate change. Missing a host could be fatal, and if parasites become extinct en masse the whole structure of ecosystems could change or even collapse.

Whether parasite or host, creeper, crawler or charismatic megafauna, with each single species a long evolutionary history is lost and a hole in an ecosystem opens up. How many holes are gaping without us even noticing? It stands to reason that most of these losses concern obscure invertebrates and therefore a menagerie of impressive slimers. With these species, their high-functioning hydrogels are gone as well. And the ones that stay might lose their grip: marine animals like mussels, starfish and slugs need biological glues to move and stay put. If that adhesive power is lost, or maybe even if it strengthens in a warmer and more acidic sea, their slime-based lifestyle is under threat. When it comes to marine mucuses and gels, more than one fragile balance is at stake.

'Three-dimensional hydrogels of organic polymers have been suggested to affect a variety of processes in the ocean, including element cycling, microbial ecology, food-web dynamics, and air-sea exchange,' states a recent publication, if only in regard to gel-like particles in the water. 'However, their abundance and distribution in the ocean are hardly known, strongly limiting an assessment of their global significance. As a consequence, marine gels are often disregarded in biogeochemical or ecosystem models.' They are being grouped according to properties like size and transparency, but these categories can be transient. Marine gels are produced and dissemble, coagulate and fragment, change their composition or become colonized by microbes, all of which might change in the new marine environments.

If that happens, the biological pump that drives organic carbon into the deep might get disrupted as well. Life on Earth is based on carbon, which gets perpetually cycled between the spheres of the planet: soil, air, stone, water and biological beings. The biological pump – or marine carbon pump – is a part of this, not as a hierarchical chain of events but the sum of interlinked processes. They all feed into the removal and transformation of atmospheric and terrestrial carbon, as well as its storage in the sea's interior and sediment. Neither of these processes nor the links between them are understood down to the last detail. It is equally unclear exactly how, when and where biological slimes are involved. Or what part they might play in derailing marine carbon-cycling in the future.

The biological pump is a complex web of interactions with a few key steps, all of them slimy. Atmospheric CO_2 crosses the biofilm interface between air and sea. Phytoplankton photosynthesize the molecules – but also microgels – in the sunlit upper layers of the ocean. That results in the production of carbon-based building blocks for hard shells and skeletons, as well as for soft body mass and sticky exopolymers. In a sense, these organic units get passed down the food chain as ever-bigger predators snatch their meals. Unconsumed phytoplankton cells will sink into the deep as part of the carbon-rich 'marine snow'. Even organic particles that stay suspended because of their small size or buoyancy might make the cut as 'flakes' when they get glued together by sticky gels. Gelatinous bodies of dead animals and gel-bound faecal pellets sink to the sea floor as well.

All the processes that feed into the biological pump help to keep it in balance, though there is some flexibility to the system. The climate crisis is a consequence of CO_2 that we humans have released into the atmosphere by burning fossil fuels. The oceans have absorbed at least a third of the anthropogenic CO_2 and phytoplankton get to convert – or 'fix' – it. But while these creatures counter our folly in part, there is still a cost. That extra CO_2 changes the chemistry of the water: it is not only becoming warmer due to the rising air temperatures, but more acidic as well. What will happen next? If phytoplankton cells take up more carbon they grow and release more extracellular substances and marine gels.

If the production of very sticky microgels gets a boost there might be more and bigger flakes of marine snow, taking the carbon export to the deep sea to a new level. On the other hand, if microbes become more active they could degrade particles more quickly than usual. Then, more organic carbon stays in the upper layers of the ocean, where it might turn into atmospheric CO_2 again. Or, as the aforementioned publication stoically states: 'These scenarios are likely to happen in a warmer and more acidic future ocean but not necessarily at the same time and place.' A similar ambivalence concerns the probable thickening of the sea-surface microlayer, which could ease or hamper exchange across the interface.

But how hyperactive will those marine microbes really be? Cosy, warm waters aside, an increasing acidity could threaten their livelihood. Microgels are an important source of food and shelter for microbes, but they might break apart more easily in a changing ocean. They could become less sticky as well, which might reduce the output of 'snowflakes'. But even functioning bio-glues pose a problem, as mentioned before: microplastic particles and fibres are widespread in the oceanic water and sediment. Sticky gels might readily attach. As a hybrid with organic material, even artificial particles could become more palatable to marine organisms and sink into the depth more easily – unless they make organic flakes more buoyant and keep them in the upper layers of the ocean.

During the last few decades, marine microgels have attracted a lot of interest in different disciplines of science. Now, researchers of all backgrounds should come together to tackle these big and urgent issues – plus a few more. Take the *Gelata*, for example, and their role in carbon-cycling. Jellyfish, ctenophores and free-living tunicates like salps have just recently been identified as extremely climate-relevant. Their gelatinous bodies incorporate extraordinary amounts of 'jelly-carbon' that gets excreted in part or makes up other waste products. Much of this ends up in the depth of the ocean, just like the dead gelatinous bodies that sink too quickly for much degradation on the way. It's a bounty of carbon export. According to some calculations, the yearly amount could be on a par with the European Union's annual carbon output.

It seems hard to imagine that gelatinous bodies and gooey waste like marine snow play such an important role, but there's even more: 'This slimy stuff actually provides the basic material for carbon stores on the ocean floor,' Dietmar Müller at the University of Sydney confirms. And he knows exactly how climate-relevant that process can be. Over the course of geological time spans, a lot of organic sediment is being stored. When that process accelerated, the extensive fixing of atmospheric carbon even let the Earth go from a greenhouse to a cooling chamber, as Müller's computer simulation shows. It reconstructs the accumulation of organic material 50 million years ago, when carbon was increasingly exported to the sea floor – and the climate cooled down accordingly. Might that happen again? It would probably be too much to hope for.

Organic sediments on the sea floor first condense into a kind of slimy mud, then mineralize over millions of years. The white cliffs of Dover are just one example of marine snow which has turned to chalky limestone – and survived in that form. According to a study by the University of Texas, these sediments provide the lubricant required by our dynamic Earth, forging a connection between life and the movement of the continental plates. These are the land masses which float on top of the viscous Earth's mantle. When they collide, mountains loom out of the Earth, providing the plates do not slide over each other. This overlapping is known as subduction. It can cause enormous friction which, according to the Texas study, can be reduced by the presence of deposited sediment.

The original gel-bound stuff like marine snow or the gelatinous bodies of animals is long gone, of course. But somehow, even in its now-mineralized form the sediment seems to keep a memory of its slimy history – or at least it behaves that way. 'Sedimentary rock is viscous,' I was told by Thorsten Becker, geophysicist and co-author of the study. 'But the immense pressure at the subduction zone has caused it to become increasingly liquid over millions of years.' Enough lubricant of that kind allows the plates to glide along with relatively little friction, so quickly that there is hardly enough time for new deposits of sediment to form. This brings the process grinding to a halt. The plates are tilting, and mountains are emerging.

When eventually, over geological time spans, this rock ultimately contributes to a new accumulation of sediment together with the organic material from the upper layers of the ocean, the lubricant will do its job and the plates will gain momentum once more. This is thanks to the new sediment that has long been mineralized and hardened into stone by the time the new cycle begins. But is it really slime-free? 'Biofilms exist there as well because microbes find plenty of sustenance in the sediment,' says microbiologist Hans-Curt Flemming. 'They can also go kilometres underground, live in extreme nutrient poverty and reduce their metabolism to such a great extent that there can be as many as 20,000 years between generations. But when they do multiply, they do so inside self-made slime.'

After all, microbes have a thing for extreme environments – as long as they can bring their protective slimes along. They will weather all changes, or might even benefit from them. Already there are mysterious pathogens causing new wasting diseases in starfish and marine mussels, making the animals disintegrate into slime. Climate change may well play a part here; the warming of the oceans is presumably encouraging rising numbers of pathogenic microbes and viruses. This is but one example where Medea is already reaching out her clammy fingers, marine microbes once again having the last word in a world thrown out of balance.

Rising temperatures, acidification and over-fishing are wiping out many species in marine habitats, decimating food webs from the top down. Algal blooms, microbial mats and shoals of jellyfish might take over and stifle the already emptied-out corners of the ocean. We still don't know how all these rapid changes will play out. Some scientists fear 'the rise of slime', while the marine biologist Daniel Pauly speaks of the possibility of an 'age of slime' – or Myxocene. The first era of a slime-dominated, boring billion years marked the calm before the cacophony of a world of many voices. But this is a man-made catastrophe. It would see our planet reduced to a deafening silence.

Afterword: Space Slimes

> *There was nothing within hearing, and nothing in sight save a vast reach of black slime . . . By the fourth evening I attained the base of the mound, which turned out to be much higher than it had appeared from a distance . . . All at once my attention was captured by a vast and singular object on the opposite slope . . . despite its enormous magnitude, and its position in an abyss which had yawned at the bottom of the sea since the world was young, I perceived beyond a doubt that the strange object was a well-shaped monolith whose massive bulk had known the workmanship and perhaps the worship of living and thinking creatures.*
>
> H.P. Lovecraft, *Dagon*

The soil is burnt red and yellowed ochre beside the deep-blue waters of the ocean. Dust roads traverse the green skin of scrub like red blood vessels. The barren coast of Western Australia is a region of colour contrasts on a continent of extremes, which repelled the first Europeans to encounter it, as the author Tim Winton writes. The intensity of their disgust and horror resounds in many Australian place names. Take Kumpupintil, for example, a mostly dried up salt lake named Lake Disappointment by the Europeans. But I'm here to visit the deadly Hamelin Pool, the innermost region of Shark Bay and separated from it by an invisible barrier: the extremely salty water of this flat pool precludes the survival of almost all life, yet slimy stromatolites have found a rare refuge here. They are the modern-day descendants of the oldest life forms on Earth, existing for at least 3.5 billion years. Hamelin Pool gives us an inkling of what the early days of life on Earth may have looked like.

Are we all alone out here? A single ear of wheat in a large field is as strange as a single world in infinite space. These were the words – or thereabouts – of the Greek philosopher Metrodorus of Chios in the fourth century BCE. Since antiquity, the question of extraterrestrial life has lost none of its urgency, becoming all the more explosive in the age of the Anthropocene. Where will we find sanctuary when this planet, ruined by our own hand, becomes too small for us, once fertile habitats are subsumed by slime? The search for greenable lands, or those with the potential to host life, elsewhere in the cosmos could become a necessity.

For the time being, however, we continue to stare into space; humanity's enquiring mind hopes to chart the final frontier and, in so doing, find its way home. 'In a cosmic setting vast and old beyond ordinary human understanding we are a little lonely,' writes Carl Sagan. 'In the deepest sense the search for extraterrestrial intelligence is a search for ourselves.' The concept of defining ourselves by that which is most alien to us also lies at the heart of the stories and cosmic horror of H.P. Lovecraft: the power struggle between monster and man. Slime often serves as a link between the two, alien yet familiar. But the superficial message endures: if we ever achieve contact with cosmic powers, we would need to be well armed.

Our technology and our strategy will determine whether such an encounter is even possible. How to locate an alien biology on as yet unreachably distant celestial bodies with mere telescopes and computer simulations? How to recognize an alien biology if it differs fundamentally from life on Earth? 'Somewhere else there might exist exotic biologies, technologies and societies,' Sagan writes. Extraterrestrial intelligence could also be far older and advanced enough to make a surprise visit of its own, leading it to our galaxy or even our solar system in order to track us down.

On 19 October 2017, the Canadian astronomer Robert Weryk observed with the Pan-STARRS telescope in Hawaii an unidentified flying object, which science has studied and puzzled over ever since. It appeared to be little more than a point of light on an arched trajectory through our solar system; the size and shape of the object could only be speculated upon. If calculations are correct, however, it was a surprisingly small, cigar-shaped object made from stone or metal. The discovery presented the Minor Planet Center, which deals with such unassuming astronomical bodies, with a problem, though: the object was as unpredictable in its classification as its seemingly chaotic tumbling trajectory through space.

Initially it was classified as comet C/2017 U1, but due to some inconsistent characteristics, such as the lack of a tail, it later became an asteroid, A/2017 U1. But this classification failed to explain the object's mysterious acceleration. Its route took it close enough to the sun to divert the small body from its previous flight path with new momentum. Instead of slowing down as its distance from the sun increased, however, the object continued to accelerate. So was it a comet after all, propelled like a rocket by the thrust of its own outgassing? We lack any visible evidence of this, such as an elongated tail or similar. The complex knot of classification seemed impossible to tease apart, but ultimately it was torn in two.

This, it transpired, was the first documented interstellar object, traversing our solar system having originated not inside it, but in the depths of space. The Minor Planet Center swiftly introduced a new classification, the I numbers, and designated 1I/ 'Oumuamua, the first and hitherto only example of this phenomenon. The name, the Minor Planet Center explained, was Hawaiian, expressing the fact that the object seemed to have been sent 'like a scout or messenger from the distant past to reach out to us'. If the idea of unknown worlds reaching out to us seems scary, then Shmuel Bialy and Abraham Loeb, Harvard astrophysicists, did nothing to offer relief. Instead, they came up with some thoroughly troubling thoughts on the matter.

"Oumuamua represents a new class of thin interstellar material, either produced naturally, through a yet unknown process . . . or of an artificial origin,' they wrote in a speculative publication. In other words, they suggest that the driving force behind the mysterious acceleration could be caused by solar radiation pressure, caught by a very thin sail less than a millimetre thick. In theory, this lightsail could arise from the disintegration of a large celestial body. Or it could have been specifically created by extraterrestrial intelligence, similar to experiments currently being carried out by scientists on Earth, Loeb among them. Could this be a deliberately dispatched scout? The latter is pure speculation, Bialy and Loeb hastened to add.

★★★

With the sun blazing above me, I walk up and down the boardwalk that overlooks the stromatolites in Hamelin Pool, modern representatives of one of the oldest life forms. But they look like stones to me, not at all spectacular. They're hiding their biological origins all too well, and the living layer of bacterial slime that covers them cannot be seen with the naked eye. I feel like I'm at the foot of a volcano, surrounded by gentle waves. The lower microbial mats close to the shore are gritty, like dark scree, while the rounded stromatolites remain underwater even at low tide, reminding me of frozen lava flows. They give no inkling of their great importance. So what are stromatolites? These ecological communities are dominated by cyanobacteria, which invented photosynthesis to pump the oceans and the atmosphere full of oxygen. They're still doing the same job in the topmost layers on their stromatolites, a gossamer-thin skin of slime that will mineralize and harden in time. Once their old home becomes uninhabitable, the cyanobacteria will do what they do best: build a new slimy layer on top. It's a painstaking business, but one that has been tried and tested over billions of years. During the unchallenged reign of microbes, stromatolites grew many metres in height and built vast reefs at the edges of the oceans.

The true nature of 'Oumuamua will probably forever remain a mystery because the object itself has long since travelled out of our reach. But perhaps the scout came not alone. We don't know if it really was our first encounter of the interstellar kind. Many cosmic objects may have visited our solar system in the past and could well be somewhere out there in the world now, unseen by us due to their diminutive size. Yet higher-powered telescopes could one day confirm Lovecraft's worst fear: that cosmic horrors walk among us, established and unseen.

We're just not quite there yet. In order to find extraterrestrial life, we will have to look for clues which correspond to some kind of biology. Yet if alien life is too exotic or too advanced in comparison to us, we may well fail to recognize it. This scenario is distantly reminiscent of the not at all communicative ocean in Stanisław Lem's *Solaris.* 'This correspondence convinced the scientists that they were confronted with a monstrous entity endowed with reason, a protoplasmic ocean-brain enveloping the entire planet,' writes Lem. 'Our instruments had intercepted minute random fragments of a prodigious and everlasting monologue unfolding in the depths of this colossal brain, which was inevitably beyond our understanding.'

It's also conceivable that extraterrestrial life is undergoing a process of development similar to ours on Earth, merely with a head start which has allowed it to throw off its biological shell. 'I think it very likely – in fact, inevitable – that biological intelligence is only a transitory phenomenon, a fleeting phase in the evolution of intelligence in the universe,' writes the astrobiologist Paul Davies in *The Eerie Silence.* 'If we ever encounter extraterrestrial intelligence, I believe it is overwhelmingly likely to be post-biological in nature.' There is room for discussion about the probability of the existence of an advanced life form of this kind. In practice, however, most astrobiologists are anticipating a much less futuristic scenario.

The benchmark for potentially inhabitable objects in space is inevitably the Earth, with its liquid water and pleasant temperature afforded by its comfortable distance from a star. A suspected salt lake beneath an ice sheet on Mars could also offer similar conditions as well, as could many millions of nooks and crannies on other celestial bodies. But what kind of life could be found there? 'I think there is a general view in astrobiology that, if life exists beyond Earth, microbial life will be more prevalent than multicellular organisms,' the NASA astrobiologist Tori Hoehler tells me.

'There are several reasons for this view,' he explains. 'One is the idea that organic life anywhere would likely have to pass through a simple, microbe-like phase before evolving into more complex multicellular life. Microbes as an overall group also grow across a greater range of environmental conditions, and with a far greater diversity of energy sources than do complex organisms. So there may be many worlds where microbes could thrive and complex life could not.'

Even today, some terrestrial species of microbes continue to demonstrate how harsh environments can be populated. They tolerate extreme temperatures, boiling acid or nuclear radiation – and are usually covered in a coating of gel. This is what makes these Earth-bound survivors so interesting to astrobiologists, whether they live in the depths of the ocean, under the ice, or in hot springs. There are even microbial communities living in the Earth's crusts as part of the deep biosphere which may well have been sealed away, inside the tiniest pores without light or contact with the Earth's surface, for millions of years. These *Subsurface lithoautotrophic microbial ecosystems* – or SLiMEs – inhabit the grey zone between life and death.

Underground caves are similarly fascinating if more accessible, coated in thick layers of biofilms. Sometimes stringy droplets of slime hang down from the roofs of the caves and are known as *snottites*. If these droplets are dripping with sulphuric acid, as is the case in one underground system of caves in the US, scientists can only enter in full protective clothing complete with breathing apparatus. The microbes, however, are not remotely concerned by this extreme environment. After all, they are embedded in a luxurious coating of slime, where they create cosy microclimates for themselves. Initial studies have shown that microbes on other planets would be able to protect themselves in much the same way.

In another experiment, terrestrial microbes were placed in water which might recreate the conditions of the underground lake on Mars. The microbes survived this salty environment, particularly when embedded in a biofilm. Does this mean there must be slimes in space? 'At the very least, the possibility of extraterrestrial slime worlds cannot be excluded, providing there is adequate water present,' says Hans-Curt Flemming, who has tested the ability of microbes to survive under extraterrestrial conditions, such as in a vacuum, in intensely dry surroundings or exposed to powerful radiation. Attached to the outside of the international space station, the particularly robust species *Deinococcus geothermalis* was able to withstand the various stresses it was exposed to for nearly two years.

There was a difference in survival rates, though: very few unicellular organisms withstood on their own, while microbes in a dried-out biofilm survived this test of endurance in large numbers. How probable is it, then, that our own terrestrial slimes could pack up and travel the universe, potentially infecting other worlds? In a sense, it would turn Lovecraft's nightmare on its head. 'Microbes would be able to survive the journey trapped inside minerals which are propelled into space by the impact of meteorites colliding with the earth,' says Flemming. 'But they wouldn't be able to leave the solar system, they could get only about as far as Jupiter.'

★ ★ ★

As I gaze out on the unassuming field of stromatolites, I wonder what drove these once-widespread communities to the edge of ruin. According to one popular theory, they were assailed by evolutionary newcomers, grazing creatures like slugs, snails and crustaceans. My visit to hostile Hamelin Pool seems to confirm this. Here, the stromatolites can grow without the threat of hungry pests, even if extremely slowly, less than a millimetre per year, with hardly a chance to repair any damages. Clear white lines cut through their field. These are trails left by camel-drawn wagons, which transported goods like wool from and to boats in the bay more than 100 years ago. It was equally long ago that the German geologist Ernst Kalkowsky gave the stromatolites their name and recognized their unique nature. He did not do this in Australia, but from fossil findings in the Harz Mountains, formed around 240 million years ago during the Permian Period, following the largest mass extinction in the Earth's history. What is the message here? Climate change could well achieve what catastrophes like this have done before and push life on Earth to the edge. But slime has always made a comeback, or never even gone away. Instead of forcing our will on the rest of nature, perhaps we could try slime's soft and slow approach for a change. Yielding instead of pushing through, a will to adapt, and even more flexibility might just enable us to survive these crises of our own making.

Further Reading

ARTICLES

Ackermann, T., 'Schönheit und Abgrund', *Weltkunst* (special issue, *Zeit* Weltkunst Verlag) (2017).

Agostini, S. et al., 'Ocean acidification drives community shifts towards simplified non-calcified habitats in a subtropical-temperate transition zone', *Scientific Reports* (2018), DOI: 10.1038/S41598-018-29251-7.

Alberti, S. et al., 'Considerations and challenges in studying liquid-liquid phase separation in biomolecular condensates', *Cell* (2019), DOI: 10.1016/j.cell.2018.12.035.

Alegado, R. and King, N: 'Bacterial influences on animal origins', *Cold Spring Harbor Perspectives in Biology* (2014), DOI: 19.1101/csh-perspect.a016162.

Allwood, C.A. et al., 'Reassessing evidence of life in 3,700-million-year-old rocks of Greenland', *Nature* (2018), DOI:10.1038/S41586-018-0610-4.

Amabebe, E. and Anumba, D.O.C., 'The vaginal micro-environment: The physiologic role of lactobacilli', *Frontiers in Medicine* (2018), DOI: 10.3389/fmed.2018.00181.

Ambort, D. et al., 'Calcium and pH-dependent packing and release of the gel-forming MUC2 mucin', *PNAS* (2012), DOI: 10.1073/pnas.1120269109.

Ambort, D. et al., 'Perspectives on mucus properties and formation – Lessons from the biochemical world', *Cold Spring Harbor Perspectives in Medicine* (2012), DOI: 10.1101/ cshperspect.a014159.

Ames, C.L. et al., 'Cassiosomes are stinging-cell structures in the mucus of the upside-down jellyfish *Cassiopea xamachana*', *Communications Biology* (2020), DOI:10.1038/s42003-020-0777-83390/jof6040234.

Aslam, S.N. et al., 'Identifying metabolic pathways for production of extracellular polymeric substances by the diatom Fragilariopsis cylindrus inhabiting sea ice', *The ISME Journal* (2018), DOI: 10.1038/s41396-017-0039-z.

Azua-Bustos, A. et al., 'Unprecedented rains decimate surface microbial communities in the hyperarid core of the Atacama Desert', *Scientific Reports* (2018), DOI: 10.1038/s41598-018-35051-w.

Bakshani, C.R. et al., 'Evolutionary conservation of the anti-microbial function of mucus: A first defence against infection', *Npj Biofilms and Microbiotas* (2018), DOI: 10.1038/s41522-018-0057-2.

Banani, S.F. et al., 'Biomolecular condensates: Organizers of cellular biochemistry', *Nature Reviews. Molecular Cell Biology* (2017), DOI: 10.1038/ nrm.2017.7.

Barnett, J.B. et al., 'Imperfect transparency and camouflage in glass frogs', *PNAS* (2020), DOI: 10.1073/pnas.1919417117.

Barr, J.J. et al., 'Bacteriophage adhering to mucus provide a non-host derived immunity', *PNAS* (2013), DOI: 10.1073/ pnas.1305923110.

Batchelor, M.T. et al., A biofilm and organomineralisation model for the growth and limiting size of ooids', *Scientific Reports* (2018), DOI: 10.1038/s41598-017-18908-4.

Baum, C. et al., 'A zymogel enhances the self-cleaning abilities of the skin of the pilot whale (*Globicephala melas*)', *Comparative Biochemistry and Physiology, Part A* (2001), DOI: 10.1016/S1095-6433(01)00445-7.

Baum, C. et al., 'Surface properties of the skin of the pilot whale *Globicephala melas*', *Biofouling* (2003), DOI: 10.1 080/0892701031000061769.

Behr, W.M. and Becker, T.W., 'Sediment control on subduction plate speeds', *Earth and Planetary Science Letters* (2018), DOI: 10.1016/ j.epsl.2018.08.057.

Belcaid, M. et al., 'Symbiotic organs shaped by distinct modes of genome evolution in cephalopods', *PNAS* (2019), DOI: 10.1073/ pnas. 1817322116.

Bernard, P. et al., 'Physics and hydraulics of the rhizo-sphere network', *Journal of Plant Nutrition and Soil Science* (2018), DOI: 10.1002/ jpln.201800042.

Bhatia, R. et al., 'Cancer-associated mucins: Role in immune modulation and metastasis', *Cancer Metastasis Reviews* (2019), DOI: 10.1007/s10555-018-09775-0.

Bialy, S. and Loeb, A., 'Could solar radiation pressure explain 'Oumuamua's peculiar acceleration?', *Astrophysical Journal Letters* (2018), DOI: 10.3847/2041-8213/aaeda8.

Bittleston, L.S. et al., 'Convergence between the microcosms of Southeast Asian and North American pitcher plants', *eLife* (2018), DOI: 10.7554/eLife.36741.

Boetius, B., 'Global change microbiology – big questions about small life for our future', *Nature Reviews Microbiology* (2019), DOI: 10.1038/s41579-019-0197-2.

Boeynaems, S. et al., 'Protein phase separation: A new phase in cell biology', *Trends in Cell Biology* (2018), DOI: 10.1016/j. tcb.2018.02.004.

Böni, L. et al., 'Hagfish slime and mucin flow properties and their implications for defense', *Scientific Reports* (2016), DOI: 10.1038/srep 30371.

Bosch, T.C., 'Rethinking the role of immunity: Lessons from Hydra', *Trends in Immunology* (2014), DOI: 10.1016/ j.it.2014.07.008.

Bouchery, T. et al., 'Hookworms evade host immunity by secreting a deoxyribonuclease to degrade neutrophil extracellular traps', *Cell Host & Microbe* (2020), DOI: 10.1016/j.chom.2020.01.011.

Bowling, A.J., 'Immunocytochemical characterization of tension wood: Gelatinous fibers contain more than just cellulose', *American Journal of Botany* (2008), DOI: 10.3732/ajb.2007368.

Brain, R.B., 'How Edvard Munch and August Strindberg contracted protoplasmania: Memory, synesthesia, and the vibratory organism in fin-de-siècle Europe', *Interdisciplinary Science Review* (2010), DOI: 10.1179/030801810 X12628670445383.

Brau, F. et al., 'Dynamics of prey prehension by chameleons

through viscous adhesion', *Nature Physics* (2016), DOI: 10.1038/nphys3795.

Brunet, T. and King, N., 'The origin of animal multicellularity and cell differentiation', *Developmental Cell* (2018), DOI: 10.1016/j.devcel.2017. 09.016.

Buckley, J. et al., 'Biparental mucus feeding: A unique example of parental care in an Amazonian cichlid', *Journal of Experimental Biology* (2010), DOI: 10.1242/jeb.042929.

Burger, A.E., 'Dispersal and germination of seeds of *Pisonia grandis*, an Indo-Pacific tropical tree associated with insular seabird colonies', *Journal of Tropical Ecology* (2005), DOI: 10.1017/SO266467404002159.

Byrd, A.L. et al., 'The human skin microbiota', *Nature Reviews. Microbiology* (2018), DOI: 10.1038/nrmicro.2017.157.

Cáliz, J. et al., 'A long-term survey unveils strong seasonal patterns in the airborne microbiota coupled to general and regional atmospheric circulations', *PNAS* (2018), DOI: 10.1073/pnas.1812826115.

Cavalier-Smith, T., 'Origin of animal multicellularity: Precursors, causes, consequences – the choanoflagellate/sponge transition, neurogenesis and the Cambrian explosion', *Philosophical Transactions of the Royal Society B. Biological Sciences* (2017), DOI: 10.1098/rstb.2015.0476.

Celli, J. et al., '*Helicobacter pylori* moves through mucus by reducing mucin viscoelasticity', *PNAS* (2009), DOI: 10.1073/pnas.0903438106.

Cepon-Robins, T.J. et al., 'Pathogen disgust sensitivity protects against infection in a high pathogen environment', *PNAS* (2021), DOI: 10.1073/pnas.2018552118.

Cerullo, A.R. et al., 'Comparative animal mucomics: Inspiration for functional materials from ubiquitous and understudied biopolymers', *ACS Biomaterials* (2020), DOI: 10.1021/acsbiomaterials.Oc00713.

Chen, W.G., 'Charge influences substrate recognition and self-assembly of hydrophobic FG sequences', *Biophysical Journal* (2017), DOI: 10.1016/j.bpj.2017.08.058.

Chen, X.D. et al., 'Hindered erosion: The biological mediation of noncohesive sediment behavior', *Water Resources Research* (2017), DOI: 10.1002/2016WR020105.

Choy, C.A., Haddock, S.H.D. and Robinson, B.H., 'Deep pelagic food web structure as revealed by in situ feeding observations', *Proceedings of the Royal Society B. Biological Sciences* (2017), DOI: 10.1098/ rspb.2017.2116.

Choy, C.A. et al., 'The vertical distribution and biological transport of marine microplastics across the epipelagic and mesopelagic water column', *Scientific Reports* (2019), DOI: 10.1038/ s41598-019-44117-2.

Cohen, E.J. et al., '*Campylobacter jejuni* motility integrates specialized cell shape, flagellar filament, and motor, to coordinate action of its opposed flagella', *PLOS Pathogens* (2020), DOI: 10.1371/journal.ppat.1008620.

Cohen, M., 'Notable aspects of glycan-protein interactions', *Biomolecules* (2015), DOI: 10.3390/biom5032056.

Cohen, S. et al., 'How viruses access the nucleus', *Biochimica et Biophysica Acta* (2011), DOI: 10.1016/j.bbamcr. 2010.12.009.

Concha, A. et al., 'Oscillation of the velvet worm slime jet by passive hydrodynamic instability', *Nature Communications* (2015), DOI: 10. 1038/ncomms7292.

Conley, K.R. et al., 'Mammoth grazers on the ocean's minuteness: A review of selective feeding using mucous meshes', *Proceedings of the Royal Society B. Biological Sciences* (2018), DOI: 10.1098/ rspb.2018.0056.

Connor, V.M., 'The use of mucous trails by intertidal limpets to enhance food resources', *The Biological Bulletin* (1986), DOI: 10.2307/1541623.

Connor, V.M. and Quinn, J.F., 'Stimulation of food species growth by limpet mucus', *Science* (1984), DOI: 10.1126/ science.225.4664.843.

Cornick, S. et al., 'Entamoeba histolytica-induced mucin exocytosis is mediated by VAMP8 and is critical in mucosal innate host defense', *mBio* (2017), DOI: 10.1128/mBio.01323-17.

Cox, K.D. et al., 'Human consumption of microplastics', *Environmental Science & Technology* (2019), DOI: 10.1021/acs.est.9b01517.

Culpepper, P.D. et al., 'Visually activating pathogen disgust: A new instrument for studying the behavioural immune system', *Frontiers in Psychology* (2018), DOI: 10.3389/fpsyg.2018.01397.

Cunliffe M. et al.: 'Sea surface microlayers: A unified physicochemical and biological perspective of the air-ocean interface', *Progress in Oceanography* (2013), DOI: 10.1016/j.pocean.2012.08.004.

Cunliffe, M. and Murrell, J.C., 'The sea-surface microlayer is a gelatinous biofilm', *The ISME Journal* (2009), DOI: 10.1038/ismej.2009.69.

Curtis, V. et al., 'Evidence that disgust evolved to protect from risk of disease', *Proceedings of the Royal Society B. Biological Sciences* (2004), DOI: 10.1098/rsbl.2003.0144.

Curtis, V. et al., 'Disgust as an adaptive system for disease avoidance behaviour', *Proceedings of the Royal Society B. Biological Sciences* (2011), DOI: 10.1098/rstb.2010.0117.

Curtis, V. and de Barra, M., 'The structure and function of pathogen disgust', *Philosophical Transaction of the Royal Society B. Biological Sciences* (2018), DOI: 10.1098/RSTB.2017.0208.

Dabiri, J.O. et al., 'Flow patterns generated by oblate medusan jellyfish: Field measurements and laboratory analyses', *Journal of Experimental Biology* (2005), DOI: 10.1242/jeb.01519.

Dadon-Pilosof, A. et al., 'Surface properties of SAR11 bacteria facilitate grazing avoidance', *Nature Micro-biology* (2017), DOI: 10.1038/s41564-017-0030-5.

Dadon-Pilosof, A. et al., 'Prey taxonomy rather than size determines salp diets', *Limnology and Oceanography* (2019), DOI: 10.1002/lno.11165.

Dassarma, P. and Dassarma, S., 'Survival of microbes in Earth's stratosphere', *Current Opinion in Microbiology* (2018), DOI: 10.1016/j.mib.2017. 11.002.

Davies, J.E. et al., 'Concise review: Wharton's jelly: The rich, but enigmatic, source of mesenchymal stromal cells', *Stem Cells Translational Medicine* (2017), DOI: 10.1002/sctm.16-0492.

Davies, M.S. and Beckwith, P., 'Role of mucus trails and trail-following in the behaviour and nutrition of the periwinkle *Littorina littorea*', *Marine Ecology Progress Series* (1999), DOI: 10.3354/meps179247.

Davies, M.S. and Blackwell, J., 'Energy saving through trail following in a marine snail', *Proceedings of the Royal Society B. Biological Science* (2007), DOI: 10.1098/rspb.2007.0046.

Davis, A.L. et al., 'Testing Darwin's hypothesis about the wonderful Venus flytrap: Marginal spikes form a "horrid prison" for moderate-sized insect prey', *American Naturalist* (2018), DOI: 10.1086/701433.

Davis-Berg, E.C., 'The predatory snail *Euglandina rosea* successfully follows mucous trails of both native and non-native prey snails', *Invertebrate Biology* (2012), DOI: 10.1111/j.1744-7410.2011.00251.x.

Deamer, D. et al., 'Hydrothermal chemistry and the origin of cellular life', *Astrobiology* (2018), DOI: 10.1089/ast.2018.1979.

De Araujo, G.G. et al., 'Survival and ice nucleation activity of *Pseudomonas syringae* strains exposed to simulated high-altitude atmospheric conditions', *Scientific Reports* (2019), DOI: 10.1038/s41598-019-44283-3.

Decho, A.W. and Gutierrez, T., 'Microbial extracellular polymeric substances (EPSs) in ocean systems', *Frontiers in Microbiology* (2017), DOI: 10.3389/fmicb.2017.00922.

Deng, J. et al., 'A bioinspired medical adhesive derived from skin secretion of *Andrias davidianus* for wound healing', *Advanced Functional Materials* (2019), DOI: 10.1002/adfm.201809110.

Denny, M.W., 'The role of gastropod pedal mucus in locomotion', *Nature* (1980), DOI: 10.1038/285160a0.

Denny, M.W., 'Mechanical properties of pedal mucus and their consequences for gastropod structure and performance', *American Zoologist* (1984), DOI: 10.1093/icb/24.1.23.

Denny, M.W., 'Invertebrate mucous secretions: Functional alternatives to vertebrate paradigms', *Symposia of the Society of Experimental Biology 43* (1989), pp. 337–66.

Denny, M.W. and Blanchette, C.A., 'Hydrodynamics, shell shape, behavior and survivorship in the owl limpet *Lottia gigantea*', *Journal of Experimental Biology 203* (2000), pp. 2623–39.

Ding, M. et al., 'Multifunctional soft machines based on stimuli-responsive hydrogels: From freestanding hydrogels to smart integrated systems', *Materialstoday Advances* (2020), DOI: 10.1016/j.mtadv.2020.100088.

Di Palma, V., 'Architecture and the organic metaphor', *The Journal of Architecture* (2006), DOI: 10.1080/13602360601037644.

Dobretsov, S. and Rittschof, D., 'Love at first taste: Induction of larval settlement by marine microbes', *International Journal of Molecular Sciences* (2020), DOI: 10.3390/ijms21030731.

Dodd, M.S. et al., 'Evidence for early life in Earth's oldest hydrothermal vent precipitates', *Nature* (2017), DOI: 10.1038/nature21377.

Driouich, A. et al., 'Root border cells and secretions as critical elements in plant host defense', *Current Opinion in Plant Biology* (2013), DOI: 10.1016/j.pbi.2013.06.010.

Driouich, A. et al., 'Root extracellular traps versus neutrophil extracellular traps in host defence, a case of functional convergence?', *Biological Reviews of the Cambridge Philosophical Society* (2019), DOI: 10.1111/brv.12522.

Drobot, B. et al., 'Compartmentalised RNA catalysis in membrane-free coacervate protocells', *Nature Communications* (2018), DOI: 10.1038/s41467-018-06072-w.

Du Plessis, E. H., 'Sartre, existentialism and panic attacks', *The Linacre Quarterly* (1992), http://epublications.marquette.edu/lnq/vol59/iss2/9.

Dutkiewicz, A. et al., 'Sequestration and subduction of deep-sea carbonate in the global ocean since the Early Cretaceous', *Geology* (2018), DOI: 10.1130/G45424.1.

Eme, L. et al., 'Archaea and the origin of eukaryotes', *Nature Reviews. Microbiology* (2017), DOI: 10.1038/nrmicro.2017.133.

Engel, A. et al., 'Marvelous marine microgels: On the distribution and impact of gel-like particles in the oceanic water-column', *frontiers in Marine Science* (2020), DOI: 10.3389/fmars.2020.00405.

Esterházy, D. et al., 'Compartmentalized gut lymph node drainage dictates adaptive immune responses', *Nature* (2019), DOI: 10.1038/s41586-019-1125-3.

Esther, C.R., 'Mucus accumulation in the lungs precedes structural changes and infection in children with cystic fibrosis', *Science Translational Medicine* (2019), DOI: 10.1126/scitranslmed.aav3488.

Ewoldt, R.H. et al., 'Rheological fingerprinting of gastropod pedal mucus and synthetic complex fluids for biomimicking adhesive locomotion', *Soft Matter* (2007), DOI: 10.1039/B615546D.

Fagherazzi, S. et al., 'Ecogeomorphology of salt marshes', *Ecogeomorphology* (2013), DOI: 10.1016/B978-0-12-374739-6.00403-6.

Fairclough, S.R. et al., 'Multicellular development in a choanoflagellate', *Current Biology* (2010), DOI: 10.1016/j.cub.2010.09.014.

Fenberg, P.B. and Roy, K., 'Anthropogenic harvesting pressure and changes in life history: Insights from a rocky intertidal limpet', *American Naturalist* (2012), DOI: 10.1086/666613.

Feuda, R. et al., 'Improved modeling of compositional heterogeneity supports sponges as sister to all other animals', *Current Biology* (2017), DOI: 10.1016/j.cub.2017.11.008.

Flemming, H.C., 'EPS – then and now', *Microorganisms* (2016), DOI: 10.3390/microorganisms4040041.

Flemming, H.C. et al., 'Biofilms: An emergent form of bacterial life', *Nature Reviews. Microbiology* (2016), DOI: 10.1038/nrmicro.2016.94.

Forget, A. et al., 'Mechanically defined microenvironment promotes stabilization of microvasculature, which correlates with the enrichment of a novel piezo−1+ population of circulating CD11b+/CD115+ monocytes', *Advanced Materials* (2019), DOI: 10.1002/adma.201808050.

Fowler, J.E. et al., 'Surface chemistry of the frog sticky-tongue mechanism', *Biointerphases* (2018), DOI: 10.1116/1.5052651.

France, M.M. and Turner, J.R., 'The mucosal barrier at a glance', *Journal of Cell Science* (2017), DOI: 10.1242/ jcs.193482.

Frantz, C., Stewart, K.M. and Weaver, V.M., 'The extracellular matrix at a glance', *Journal of Cell Science* (2010), DOI: 10.1242/ jcs.023820.

Freckelton, M.L. et al., 'Induction of invertebrate larval settlement: Different bacteria, different mechanisms?', *Scientific Reports* (2017), DOI: 10.1038/srep42557.

Freedman, B.R. et al., 'The (dys)functional extracellular matrix', *Biochimica et Biophysica Acta* (2015), DOI: 10.1016/j.bbamcr.2015.04.015.

Frenkel, E.S. and Ribbeck, K., 'Salivary mucins in host defense and disease prevention', *Journal of Oral Microbiology* (2015), DOI: 10.3402/jom.ccv7.29759.

Frenkel, E.S. and Ribbeck, K., 'Salivary mucins protect surfaces from colonization by cariogenic bacteria', *Applied and Environmental Microbiology* (2015), DOI: 10.1128/AEM.02573-14.

Frey, S. et al., 'Surface properties determining passage rates of proteins through nuclear pores', *Cell* (2018), DOI: 10.1016/j.cell.2018.05.045.

Frösler, J. et al., 'Survival of *Deinococcus geothermalis* in biofilms under desiccation and simulated space and Martian conditions', *Astrobiology* (2017), DOI: 10.1089/ast.2015.1431.

Fudge, D.S. et al., 'Composition, morphology, and mechanics of hagfish slime', *Journal of Experimental Biology* (2005), DOI: 10.1242/ jeb.01963.

Fudge, D.S. et al., 'Hagfish slime threads as a biomimetic model for high performance protein fibres', *Bioinspiration & Biomimetics* (2010), DOI: 10.1088/1748-3182/5/3/035002.

Furter, M. et al., 'Mucus architecture and near-surface swimming affect distinct *Salmonella typhimurium* infection patterns along the murine intestinal tract', *Cell* (2019), DOI: 10.1016/j.celrep.2019.04.106.

Galgani, L. et al., 'Effects of ocean acidification on the biogenic composition of the sea-surface microlayer: Results from a mesocosm study', *JGR Oceans* (2014), DOI: 10.1002/2014JC010188.

Gao, C. et al., 'Single-cell bacterial transcription measurements reveal the importance of dimethylsulfoniopropionate (DMSP) hotspots in ocean sulfur cycling', *Nature Communications* (2020), DOI: 10.1038/s41467-020-15693-z.

Geerlings, S.Y. et al., '*Akkermansia muciniphila* in the human gastrointestinal tract: When, where, and how?', *Microorganisms* (2018), DOI: 10.3390/microorganisms6030075.

Gemmell, B.J. et al., 'The most efficient metazoan swimmer creates a "virtual wall" to enhance performance', *Proceedings of the Royal Society B. Biological Sciences* (2021), DOI: 10.1098/rspb.2020.2494.

Gerringer, M.E. et al., 'Distribution, composition and functions of gelatinous tissues in deep-sea fishes', *The Royal Society Publishing* (2017), DOI: 10.1098/rsos.171063.

Gerringer, M.E. et al., '*Pseudoliparis swirei* sp. nov.: A newly-discovered hadal snailfish (*Scorpaeniformes: Liparidae*) from the Mariana Trench', *Zootaxa* (2017), DOI: 10.11646/zootaxa.4358.1.7.

Gilet, T. and Bourouiba, L., 'Rain-induced ejection of pathogens from leaves: Revisiting the hypothesis of splash-on-film using high-speed visualization', *Integrative & Comparative Biology* (2014), DOI: 10.1093/icb/icu116.

Gilet, T. and Bourouiba, L., 'Fluid fragmentation shapes rain-induced foliar disease transmission', *Journal of the Royal Society Interface* (2015), DOI: 10.1098/rsif.2014.1092.

Gillois, K. et al., 'Mucus: An underestimated gut target for environmental pollutants and food additives', *Microorganisms* (2018), DOI: 10.3390/microorganisms6020053.

Goldmann, O. and Medina, E., 'The expanding world of extracellular traps: Not only neutrophils but much more', *Frontiers in Immunology* (2013), DOI: 10.3389/fimmu.2012.00420.

Goldstein, M.C. et al., 'Increased oceanic microplastic debris enhances oviposition in an endemic pelagic insect', *Biology Letters* (2012), DOI: 10.1098/rsbl.2012.0298.

Gomes, E. and Shorter, J., 'The molecular language of membraneless organelles', *Journal of Biological Chemistry* (2019), DOI: 10.1074/jbc.TM118.001192.

Gould, J. et al., 'Adhesive defence mucus secretions in the red triangle slug (*Triboniophorus graeffei*) can incapacitate adult frogs', *bioRxiv* (2019), DOI: 10.1101/544775.

Gowda, D.C. and Davidson, E.A., 'Structural features of carbohydrate moieties in snake venom glycoproteins', *Biochemical and Biophysical Research Communications* (1992), DOI: 10.1016/S0006-291X(05)80144-5.

Gowda, D.C. and Davidson, E.A., 'Isolation and characterization of novel mucin-like glycoproteins from cobra venom', *Journal of Biological Chemistry* 31/269 (1994), pp. 20031–9.

Grogan, L.F. et al., 'Immunological aspects of chytridiomycosis', *Journal of Fungi* (2020), DOI: 10.3390/jof6040234.

Gutow, L. et al., 'Gastropod pedal mucus retains microplastics and promotes the uptake of particles by marine periwinkles', *Environmental Pollution* (2019), DOI: 10.1016/j.envpol.2018.12.097.

Haddock, S.H.D., 'A golden age of gelata: Past and future research on planktonic ctenophores and cnidarians', *Hydrobiologia* (2004), DOI: 10.1007/s10750-004-2653-9.

Hansma, H.G., 'Better than membranes at the origin of life?', *life* (2017), DOI: 10.3390/life7020028.

Hansson, G.C. et al., 'The inner of the two Muc2 mucin-dependent mucus layers in colon is devoid of bacteria', *Gut Microbes* (2010), DOI: 10.4161/gmic.1.1.10470.

Hargens, A.R. and Shabica, S.V., 'Protection against lethal freezing temperatures by mucus in an antarctic limpet', *Cryobiology* (1973), DOI: 10.1016/0011-2240(73)90052-7.

Hauschke, N. et al., 'Ernst Kalkowsky (1851–1938): "'Oolith und Stromatolith im norddeutschen Buntsandstein". Nach 100 Jahren Nationaler Geotop: Der Heeseberg', *SDGG 63* (2009), p. 211.

Hay, P., 'Bacterial vaginosis', *F1000Research* (2017), DOI: 10.12688/f1000research.11417.1.

Hayashi, K. et al., 'Fast-forming hydrogel with ultralow polymeric content as an artificial vitreous body', *Nature Biomedical Engineering* (2017), DOI: 10.1038/s41551-017-0044.

Hays, G.C., Doyle, T.K. and Houghton, J.D.R., 'A paradigm shift in the trophic importance of jellyfish?', *Trends in Ecology & Evolution* (2018), DOI: 10.1016/j.tree.2018.09.001.

Hennig, R., 'Liegen der Erzählung vom "Geronnenen Meer" geographische Tatsachen zugrunde?', *Geographische Zeitschrift* 32/2 (1926), pp. 62–73.

Herr, J.E. et al., 'Stabilization and swelling of hagfish slime mucin vesicles', *Journal of Experimental Biology* (2010), DOI: 10.1242/jeb.038992.

Hirano, S.S. and Upper, C.D., 'Bacteria in the leaf ecosystem with emphasis on pseudomonas syringae – a pathogen, ice nucleus, and epiphyte', *Microbiology and Molecular Biology Reviews* (2000), DOI: 10.1128/MMBR.64.3.624-653.2000.

Hollingsworth, M.A. and Swanson, B.J., 'Mucins in cancer: Protection and control of the cell surface', *Nature Reviews Cancer* (2004), DOI: 10.1038/nrc1251.

Hossfeld, U. and Levit, G.S., '"Tree of life" took root 150 years ago', *Nature* (2016), DOI: 10.1038/540038a.

Hsu, B.B. et al., 'Dynamic modulation of the gut microbiota and metabolome by bacteriophages in a mouse model', *Cell Host & Microbe* (2019), DOI: 10.1016/j.chom.2019.05.001.

Hsue, Y.P. et al., 'Nematode-trapping fungi eavesdrop on nematode pheromones', *Current Biology* (2013), DOI: 10.1016/j.cub.2012.11.035.

Huang, Y. et al., 'Sundew adhesive: A naturally occurring hydrogel', *Journal of the Royal Society Interface* (2015), DOI: 10.1098/rsif.2015.0226.

Huertas, V. et al., 'Mucus-secreting lips offer protection to suction-feeding corallivorous fishes', *Current Biology* (2017), DOI: 10.1016/j.cub.2017.04.056.

Huertas, V. and Bellwood, D.R., 'Trophic separation in planktivorous reef fishes: A new role for mucus?', *Oecologia* (2020), DOI: 10.1007/s00442-020-04608-w.

Hussey, G.S. et al., 'Extracellular matrix-based materials for regenerative medicine', *Nature Reviews Materials* (2018), DOI: 10.1038/s41578-018-0023-x.

Ionov, L., 'Hydrogel-based actuators: Possibilities and limitations', *Materials Today* (2014), DOI: 10.1016/j.mattod.2014.07.002.

Jackson, J.B.C., 'Ecological extinction and evolution in the brave new ocean', *PNAS* (2008), DOI: 10.1073/pnas.0802812105.

Jardine, C.B. et al., 'Biofilm consumption and variable diet composition of Western sandpipers (*Calidris mauri*) during migratory stopover', *PLOS One* (2015), DOI: 10.1371/journal.pone.0124164.

Jared, C. et al., 'Skin gland concentrations adapted to different evolutionary pressures in the head and posterior regions of the caecilian *Siphonops annulatus*', *Scientific Reports* (2018), DOI: 10.1038/s41598-018-22005-5.

Johannesson, K. et al., 'Indiscriminate males: Behaviour of a marine snail compromised by a sexual conflict?', *PLOS One* (2010), DOI: 10.1371/journal.pone.0012005.

Johansson, M.E.V. et al., 'The gastrointestinal mucus system in health and disease', *Nature Reviews. Gastroenterology & Hepatology* (2013), DOI: 10.1038/nrgastro.2013.35.

Johansson, M.E.V. et al., 'Bacteria penetrate the normally impenetrable inner colon mucus layer in both murine colitis models and patients with ulcerative colitis', *Gut* (2014), DOI: 10.1136/gutjnl-2012-303207.

Johansson, M.E.V. and Hansson, G.C., 'Is the intestinal goblet cell a major immune cell?', *Cell Host Microbe* (2014), DOI: 10.1016/j.chom.2014.02.014.

Johansson, M.E.V. and Hansson, G.C., 'Mucus and the goblet cell', *Digestive Diseases* (2015), DOI: 10.1159/000354683.

Johansson, M.E.V. and Hansson, G.C., 'Immunological aspects of intestinal mucus and mucins', *Nature Reviews. Immunology* (2016), DOI: 10.1038/ nri.2016.88.

Jones, S.J., 'Goo, glue, and grain binding: Importance of biofilms for diagenesis in sandstones', *Geology* (2017), DOI: 10.1130/focus102017.1.

Joung, Y.S. and Buie, C.R., 'Aerosol generation by raindrop impact on soil', *Nature Communications* (2015), DOI: 10.1038/ncomms7083.

Kang, D.W. et al., 'Long-term benefit of microbiota transfer therapy on autism symptoms and gut micro-biota', *Scientific Reports* (2019), DOI: 10.1038/s41598-019421830.

Katija, K. et al., 'From the surface to the seafloor: How giant larvaceans transport microplastics into the deep sea', *Science Advances* (2017), DOI: 10.1126/sciadv.1700715.

Katija, K. et al., 'New technology reveals the role of giant larvaceans in oceanic carbon cycling', *Science Advances* (2017), DOI: 10.1126/sciadv.1602374.

Knee, E.M. et al., 'Root mucilage from pea and its utilization by rhizosphere bacteria as a sole carbon source', *Molecular Plant-Microbe Interactions* (2007), DOI: 10.1094/MPMI.2001.14.6.775.

Knoop, K.A. and Newberry, R.D., 'Goblet cells: Multi-faceted players in immunity at mucosal surfaces', *Mucosal Immunology* (2018), DOI: 10.1038/s41385-018-0039-y.

Krembs, C., Eicken, H. and Deming, J.W., 'Exopolymer alteration of physical properties of sea ice and implications for ice habitability and biogeochemistry in a warmer Arctic', *PNAS* (2011), DOI: 10.1073/pnas.1100701108.

Kufe, D.W., 'Mucins in cancer: Function, prognosis and therapy',

Nature Reviews Cancer (2010), DOI: 10.1038/nrc2761.

Kufe, D.W., 'MUC1-C oncoprotein as a target in breast cancer: Activation of signaling pathways and therapeutic approaches', *Oncogene* (2013), DOI: 10.1038/onc.2012.158.

Kuo, J.C. et al., 'Physical biology of the cancer cell glycocalyx', *Nature Physics* (2018), DOI: 10.1038/s41567-018-0186-9.

Labokha, A.A. and Fassati, A., 'Viruses challenge selectivity barrier of nuclear pores', *Viruses* (2013), DOI: 10.3390/v5102410.

Laird, M.K. et al., 'Facultative oviparity in a viviparous skink (*Saiphos equalis*)', *Biology Letters* (2019), DOI: 10.1098/rsbl.2018.0827.

Lang, T. et al., 'Gel-forming mucins appeared early in metazoan evolution', *PNAS* (2007), DOI: 10.1073/10.1073/pnas.0705984104.

Lang, T. et al., 'Searching the evolutionary origin of epithelial mucus protein components – mucins and FCGBP', *Molecular Biology and Evolution* (2016), DOI: 10.1093/molbev/msw066.

Lange, J. et al., 'Husbandry of jellyfish, from the beginning until today – Überblick über die Quallenhaltung von den Anfängen bis heute', *Der Zoologische Garten* (2016), DOI: 10.1016/j.zoolgart.2015.09.009.

Lareen, A. et al., 'Plant root-microbe communication in shaping root microbiotas', *Plant Molecular Biology* (2016), DOI: 10.1007/s11103-015-0417-8.

Laundon, D. et al., 'The architecture of cell differentiation in choanoflagellates and sponge choanocytes', *PLOS Biology* (2019), DOI: 10.1371/journal.pbio.3000226.

Lazow, S.P. et al., 'A novel two-component, expandable bioadhesive for exposed defect coverage: Applicability to prenatal procedures', *Journal of Pediatric Surgery* (2021), DOI: 10.1016/j.jpedsurg.2020.09.030.

Lee, J.W. et al., 'Hepatocytes direct the formation of a pro-metastatic niche in the liver', *Nature* (2019), DOI: 10.1038/s41586-019-1004-y.

Levin, T.C. et al., 'The rosetteless gene controls development in the choanoflagellate S. Rosetta', *eLife* (2014), DOI: 10.7554/eLife.04070.

Li, J., 'Tough adhesives for diverse wet surfaces', *Science* (2017), DOI: 10.1126/science.aah6362.

Li, L.D. et al., 'Spatial configuration and composition of charge modulates transport into a mucin hydrogel barrier', *Biophysical Journal* (2013), DOI: 10.1016/j.bpj.2013.07.050.

Lichtman, I.D. et al., 'Bedform migration in a mixed sand and cohesive clay intertidal environment and implications for bed material transport predictions', *Geomorphology* (2018), DOI: 10.31223/osf.io/cys39.

Lidell, M.E. et al., 'Entamoeba histolytica cysteine proteases cleave the MUC2 mucin in its C-terminal domain and dissolve the protective colonic mucus gel', *PNAS* (2006), DOI: 10.1073/pnas.0600623103.

Lieleg, O. and Ribbeck, K., 'Biological hydrogels as selective diffusion barriers', *Trends in Cell Biology* (2011), DOI: 10.1016/j.tcb.2011.06.002.

Linden, S.K. et al., 'Mucins in the mucosal barrier to infection', *Mucosal Immunity* (2008), DOI: 10.1038mi.2008.5.

Litsios, S., 'Charles Dickens and the movement for sanitary reform', *Perspectives in Biology and Medicine* (2003), DOI: 10.1353/pbm.2003.0025.

Liu, K. et al., 'How carnivorous fungi use three-celled constricting rings to trap nematodes', *Protein Cell* (2012), DOI: 10.1007/s13238-012-2031-8.

Lopresit, E.F. et al., 'Mucilage-bound sand reduces seed predation by ants but not by reducing apparency: A field test of 53 plant species', *Ecology* (2019), DOI: 10.1002/ecy.2809.

Madsen, S.S. et al., 'Sexual maturation and changes in water and salt transport components in the kidney and intestine of three-spined stickleback (*Gasterosteus aculeatus L.*)', *Comparative Biochemistry and Physiology Part A, Molecular & Integrative Physicology* (2018), DOI: 10.1016/j.cbpa.2015.06.021.

Mailho-Fontana, P.L. et al., 'Morphological evidence for an oral venom system in caecilian amphibians', *iScience* (2020), DOI: 10.1016/j.isci.2020.101234.

Maixner, F. et al., 'The 5,300-year-old Helicobacter pylorigenome of the Iceman', *Science* (2016), DOI: 10.1126/science. aad2545.

Malarkey, J. et al., 'The pervasive role of biological cohesion in bedform development', *Nature Communications* (2015), DOI: 10.1038/ncomms7257.

Mammola, S. et al., 'Towards a taxonomically unbiased European Union biodiversity strategy for 2030', *Proceedings of the Royal Society B. Biological Sciences* (2020), DOI: 10.1098/rspb.2020.2166.

McFall-Ngai, M. et al., 'Animals in a bacterial world, a new imperative for the life sciences', *PNAS* (2013), DOI: 10.1073/pnas.1218525110.

McKee, A. et al., 'Substrate attributes determine gait in a terrestrial gastropod', *The Biological Bulletin* (2013), DOI: 10.1086/BBLv224n1p53.

McNear, D.H., 'The rhizosphere – roots, soil and everything in between', *Nature Education Knowledge* (2013).

Mehta, S.D., 'The microbiome composition of a man's penis predicts incident bacterial vaginosis in his female sex partner with high accuracy', *frontiers in Cellular and Infection Microbiology* (2020), DOI: 10.3389/fcimb.2020.00433.

Meier, F.C. and Lindbergh, C.A., 'Collecting micro-organisms from the Arctic atmosphere', *The Scientific Monthly* 1/40 (1935), pp.5–20.

Mellerowicz, E.J. and Gorshkova, T.A., 'Tensional stress generation in gelatinous fibres: A review and possible mechanism based on cell-wall structure and composition', *Journal of Experimental Botany* (2012), DOI: 10.1093/jxb/err339.

Meyer, Justin R., 'Sticky bacteriophage protect animal cells', *PNAS* (2013), DOI: 10.1073/pnas.1307782110.

Miyashita, T. et al., 'Hagfish from the cretaceous Tethys Sea and a reconciliation of the morphological-molecular conflict in early vertebrate phylogeny', *PNAS* (2019), DOI: 10.1073/pnas.1814794116.

Möckl, L., 'The emerging role of the mammalian glycocalyx in functional membrane organization and immune system regulation', *frontiers in Cell and Developmental Biology* (2020), DOI: 10.3389/fcell.2020.00253.

Mommer, L. et al., 'Advances in the rhizosphere: Stretching the interface of life', *Plant Soil* 407/1-8 (2016), DOI: 10.1007/s11104-016-3040-9.

Moran, M.A. et al., 'Deciphering ocean carbon in a changing world', *PNAS* (2016), DOI: 10.1073/pnas.1514645113.

Morasch, M. et al., 'Heat-flow-driven oligonucleotide gelation separates single-base differences', *Angewandte Chemie Internationale Edition* (2016), DOI: 10.1002/anie.201601886.

Morris, C.E. et al., 'The life history of the plant pathogen *Pseudomonas syringae* is linked to the water cycle', *The ISME Journal* (2008), DOI: 10.1038/ismej.2007.113.

Mouw, J.K. et al., 'Extracellular matrix assembly: A multiscale deconstruction', *Nature Reviews. Molecular Cell Biology* (2014), DOI: 10.1038/nrm3902.

Mukherjee, I. et al., 'The Boring Billion: A slingshot for complex life on Earth', *Scientific Reports* (2018), DOI: 10.1038/s41598-018-22695-x.

Mustaffa, N.I.H. et al., 'High-resolution observations on enrichment processes in the sea-surface microlayer', *Scientific Reports* (2018), DOI: 10.1038/s41598-018-31465-8.

Mustaffa, N.I.H. et al., 'Global reduction of *in situ* CO2 transfer velocity by natural surfactants in the sea-surface microlayer', *Proceedings of the Royal Society A. Mathematical, Physical and Engineering Sciences* (2020), DOI: 10.1098/rspa.2019.0763

Nakashima, K. et al., 'Chitin-based barrier immunity and its predated mucus-colonization by indigenous gut microbiota', *Nature communications* (2018), DOI: 10.1038/s41467-018-05884-0.

Neumann, A. et al., 'Extracellular traps: An ancient weapon of multiple kingdoms', *Biology* (2020), DOI: 10.3390/biology9020034.

Ng, T.P.T. et al., 'Mucus trail following as a mate-searching strategy in mangrove littorinid snails', *Animal Behaviour* (2011), DOI: 10.1016/j. anbehav.2011.05.017.

Ng, T.P.T. et al., 'Snails and their trails: The multiple functions of trail-following in gastropods', *Biological Review* (2013), DOI: 10.1111/brv.12023.

Nguyen, T.A. et al., 'Fungal wound healing through instantaneous protoplasmic gelation', *Current Biology* (2021), DOI: 10.1016/j. cub.2020.10.016.

Noel, A.C. et al., 'Frogs use a viscoelastic tongue and non-Newtonian saliva to catch prey', *Journal of the Royal Society Interface* (2017), DOI: 10.1098/rsif.2016.0764.

Noor, N. et al., '3D printing of personalized thick and perfusable cardiac patches and hearts', *Advanced Science* (2019), DOI: 10.1002/ advs.201900344.

Ochs, M. et al., 'On top of the alveolar epithelium: Surfactant and the glycocalyx', *International Journal of Molecular Sciences* (2020), DOI: 10.3390/ijms21093075.

O'Hanlon, S.J. et al., 'Recent Asian origin of chytrid fungi causing global amphibian declines', *Science* (2018), DOI: 10.1126/ science.aar1965.

O'Keefe, S.J.D. et al., 'Fat, fiber and cancer risk in African Americans and rural Africans', *Nature Communications* (2015), DOI: 10.1038/ncomms7342.

Okumura, R. and Takeda, K., 'Maintenance of intestinal homeostasis by mucosal barriers', *Inflammation and Regeneration* (2018), DOI: 10.1186/s41232-018-0063-z.

Orosei, R. et al., 'Radar evidence of subglacial liquid water on Mars', *Science* (2018), DOI: 10.1126/science.aar7268.

Osaki, H. and Tagawa, K., 'Life on a deadly trap: *Buckleria paludum*, a specialist herbivore of carnivorous sundew plants, licks mucilage from glands for defense', *Entomological Science* (2020), DOI: 10.1111/ ens.12419.

Padillo-Gamino, J.L. et al., 'Formation and structural organization of the egg-sperm bundle of the scleractinian coral *Montipora capitata*', *Coral Reefs* (2011), DOI: 10.1007/s00338-010-0700-8.

Panitz, C. et al., 'Tolerances of *Deinococcus geothermalis* biofilms and planktonic cells exposed to space and simulated Martian conditions in low earth orbit for almost two years', *Astrobiology* (2019), DOI: 10.1089/ast.2018.1913.

Panova, M. et al., 'Extreme female promiscuity in a non-social invertebrate species', *PLOS One* (2010), DOI: 10. 1371/journal. pone.0009640.

Párraga, J. et al., 'Intrusions of dust and iberulites in Granada basin (Southern Iberian Peninsula). Genesis and formation of atmospheric iberulites', *Atmospheric Research* (2021), DOI: 10.1016/j. atmosres.2020.105260.

Parsons, D.R., 'The role of biophysical cohesion on subaqueous bed form size', *Geophysical Research Letters* (2016), DOI: 10.1002/2016GL067667.

Patel, K. et al., 'Mucus trail tracking in a predatory snail: Olfactory processing retooled to serve a novel sensory modality', *Brain Behavior* (2014), DOI: 10.1002/brb3.198.

Pavia, F.J. et al., 'Shallow particulate organic carbon regeneration in the South Pacific Ocean', *PNAS* (2019), DOI: 10.1073/ pnas.1901863116.

Pawlicki, J.M. et al., 'The effect of molluscan glue proteins on gel mechanics', *The Journal of Experimental Biology* (2004), DOI: 10.1242/jeb.00859.

Peixoto, A. et al., 'Protein glycosylation and tumor micro-environment alterations driving cancer hallmarks', *Frontiers in Oncology* (2019), DOI: 10.3389/fonc.2019.00380.

Pelaseyed, T. et al., 'The mucus and mucins of the goblet cells and enterocytes provide the first defense line of the gastrointestinal tract and interact with the immune system', *Immunological Reviews* (2014), DOI: 10.1111/ imr.12182.

Pereira, R. et al., 'Reduced air–sea CO2 exchange in the Atlantic

Ocean due to biological surfactants', *Nature Geoscience* (2018), DOI: 10.1038/ s41561-018-0136-2.

Philips, C.R. et al., 'Fifty years of cereal leaf beetle in the US: An update on its biology, management, and current research', *Journal of Integrated Pest Management* (2011), DOI: 10.1603/IPM11014.

Phillips, B. and Shine, R., 'When dinner is dangerous: Toxic frogs elicit species-specific responses from a generalist snake predator', *The American Naturalist* (2007), DOI: 10.1086/522845.

Pickard, J.M. et al., 'Gut microbiota: Role in pathogen colonization, immune responses and inflammatory disease', *Immunological Reviews* (2017), DOI: 10.1111/ imr.12567.

Pietsch, R.B. et al., 'Wind-driven spume droplet production and the transport of *Pseudomonas syringae* from aquatic environments', *PeerJournal* (2018), DOI: 10.7717/peerj.5663.

Pinget, G. et al., 'Impact of the food additive titanium dioxide (E171) on gut microbiota–host interaction', *Frontiers in Nutrition* (2019), DOI: 10.3389/fnut.2019.00057.

Piontek, J. et al., 'Effects of rising temperature on the formation and microbial degradation of marine diatom aggregates', *Aquatic Microbial Ecology* (2009), DOI: 10.3354/ame01273.

Pisani, D. et al.: 'Genomic data do not support comb jellies as the sister group to all other animals', *PNAS* (2015), DOI: 10.1073/ pnas.1518127112.

Pita, L. et al., 'The sponge holobiont in a changing ocean: From microbes to eco-systems', *Microbiota* (2018), DOI: 10.1186/s40168-018-0428-1.

Proctor, R., 'Architecture from the cell-soul: René Binet and Ernst Haeckel', *The Journal of Architecture* (2006), DOI: 10.1080/1360236060-1037818.

Purcell, S.C. and Godula, K., 'Synthetic glycoscapes: Addressing the structural and functional complexity of the glycocalyx', *Interface Focus* (2019), DOI: 10.1098/rsfs.2018.0080.

Rahfeld, P. et al., 'An enzymatic pathway in the human gut microbiota that converts A to universal O type blood', *Nature*

Microbiology (2019), DOI: 10.1038/s41564-019-0469-7.

Rahlff, J., 'The virioneuston: A review on viral–bacterial associations at air–water interfaces', *Viruses* (2019), DOI: 10.3390/v11020191.

Raverty, S.A. et al., 'Respiratory microbiota of endangered southern resident killer whales and microbiota of surrounding sea surface microlayer in the Eastern North Pacific', *Scientific Reports* (2017), DOI: 10.1038/ s41598-017-00457-5.

Rehbock, P.F., 'Huxley, Haeckel, and the oceanographers: The case of *Bathybius haeckelii*', *Isis* 4/66 (1975), pp.504–33.

Rice, A.L., 'Thomas Henry Huxley and the strange case of *Bathybius haeckelii*: A possible alternative explanation', *Archives of Natural History* (1983), DOI: 10.3366/anh.1983.11.2.169.

Rix, L. et al., 'Coral mucus fuels the sponge loop in warm- and cold-water coral reef ecosystems', *Science Reports* (2016), DOI: 10.1038/srep18715.

Rix, L., 'Reef sponges facilitate the transfer of coral-derived organic matter to their associated fauna via the sponge loop', *Marine Ecology Progress Series* (2018), DOI: 10.3354/meps12443.

Roberts, H.R. et al., 'Evidence for facultative protocarnivory in *Capsella bursa-pastoris* seeds', *Scientific Reports* (2018), DOI: 10.1038/s41598-018-28564-x.

Robinson, T.B. et al., 'Rising bubbles enhance the gelatinous nature of the air–sea interface', *Limnology and Oceanography* (2019), DOI: 10.1002/lno.11188.

Robison, B.H. et al., 'Giant larvacean houses: Rapid carbon transport to the deep sea floor', *Science* (2005), DOI: 10.1126/science.1109104.

Rodriguez-Caballero, E. et al., 'Dryland photoautotrophic soil surface communities endangered by global change', *Nature Geoscience* (2018), DOI: 10.1038/s41561-018-0072-1.

Rodriguez-Cabellero, E. et al., 'Ecosystem services provided by biocrusts: From ecosystem functions to social values', *Journal of Arid Environments* (2018), DOI: 10.1016/j.jaridenv.2017.09.005.

Rouse, G.W. et al., 'An inordinate fondness for *Osedax* (Siboglinidae: Annelida): Fourteen new species of bone worms from California', *Zootaxa* (2018), DOI: 10.11646/zootaxa.4377.4.1.

Ryan, J.F. et al., 'The genome of the ctenophore *Mnemiopsis leidyi* and its implications for cell type evolution', *Science* (2013), DOI: 10.1126/science.1242592.

Sarabian, C. et al., 'Evolution of pathogen and parasite avoidance behaviours', *Philosophical Transactions of the Royal Society B. Biological Sciences* (2018), DOI: 10.1098/rstb.2017.0256.

Sarabian, C. et al., 'Feeding decisions under contamination risk in bonobos', *Philosophical Transactions of the Royal Society B. Biological Sciences* (2018), DOI: 10.1098/rstb.2017.0195.

Sasse, J. et al., 'Feed your friends: Do plant exudates shape the root microbiota?', *Trends in Plant Science* (2018), DOI: 10.1016/j.tplants.2017.09.003.

Sauer, K., 'The war on slime', *Scientific American* (2017), DOI: 10.1038/scientificamerican1117-64.

Scheele, B.C. et al., 'Amphibian fungal panzootic causes catastrophic and ongoing loss of biodiversity', *Science* (2019), DOI: 10.1126/science.aav0379.

Schröder, K. et al., 'The origin of mucosal immunity: Lessons from the holobiont *Hydra*', *mBio* (2016), DOI: 10.1128/mBio.01184-16.

Schroeder, B.O., 'Fight them or feed them: How the intestinal mucus layer manages the gut microbiota', *Gastroenterology Report* (2019), DOI: 10.1093/gastro/ goy052.

Schuerch, M. et al., 'Future response of global coastal wetlands to sea-level rise', *Nature* (2018), DOI: 10.1038/ s41586-018-0476-5.

Schulze-Makuch, D. et al., 'Transitory microbial habitat in the hyperarid Atacama Desert', *PNAS* (2018), DOI: 10.1073/pnas.1714341115.

Seear, P.J. et al., 'The molecular evolution of spiggin nesting glue in sticklebacks', *Molecular Ecology* (2015), DOI: 10.1111/mec.13317.

Sharon, G. et al., 'Human gut microbiota from autism spectrum disorder promote behavioral symptoms in mice', *Cell* (2019), DOI: 10.1016/ j.cell.2019.05.004.

Shkoporov, A.N. et al., 'Bacteriophages of the human gut: The "known unknown" of the microbiota', *Cell Host and Microbe* (2019), DOI: 10.1016/j.chom.2019.01.017.

Shniukova, E.I. and Zolotareva, E.K., 'Diatom exopolysaccharides: A review', *International Journal on Algae* (2015), DOI: 10.1615/Inter JAlgae.v17.i1.50.

Sleeper, H.L. et al., 'Alarm pheromones from the marine opisthobranch *Navanax inermis*', *Journal of Chemical Ecology* (1980), DOI: 10.1007/ bf00987527.

Smith, A.M., 'The structure and function of adhesive gels from invertebrates', *Integrative & Comparative Biology* (2002), DOI: 10.1093/icb/42.6.1164.

Smith, A.M. and Morin, M.C., 'Biochemical differences between trail mucus and adhesive mucus from marsh periwinkle snails', *The Biological Bulletin* (2002), DOI: 10.2307/1543576.

Smith-Dupont, K.B. et al., 'Probing the potential of mucus permeability to signify preterm birth risk', *Scientific Reports* (2017), DOI: 10.1038/ s41598-017-08057-z.

Solé, R. et al., 'Liquid brains, solid brains', *Philosophical Transactions of the Royal Society B. Biological Sciences* (2019), DOI: 10.1098/ rstb.2019.0040.

Sonnenburg, E.D. and Sonnenburg, J.L., 'The ancestral and industrialized gut microbiota and implications for human health', *Nature Reviews Microbiology* (2019), DOI: 10.1038/s41579-019-0191-8.

Sovran, B. et al., 'Age-associated impairment of the mucus barrier function is associated with profound changes in microbiota and immunity', *Scientific Reports* (2019), DOI: 10.1038/s41598-018-35228-3.

Stahl, M. et al., 'The helical shape of *Campylobacter jejuni* promotes in vivo pathogenesis by aiding transit through intestinal mucus and colonization of crypts', *Infection and Immunity* (2016), DOI: 10.1128/ IAI.00751-16.

Stinson, L.F. et al., 'The not-so-sterile womb: Evidence that the human fetus is exposed to bacteria prior to birth', *Frontiers in Microbiology* (2019), DOI: 10.3389/fmicb.2019.01124.

Suzuki, Y. et al., 'Deep microbial proliferation at the basalt interface in 33.5–104 million-year-old oceanic crust', *Communications Biology* (2020), DOI: 10.1038/s42003-020-0860-1.

Svanberg, I., 'Black slugs (*Arion ater*) as grease: A case study of technical use of gastropods in pre-industrial Sweden', *Journal of Ethnobiology* (2009), DOI: 10.2993/0278-0771(2006)26[299:BSAAAG]2.0.CO;2.

Tanentzap, A.J. et al., 'Ungulate saliva inhibits a grass-endophyte mutualism', *Biology Letters* (2014), DOI: 10.1098/rsbl.2014.0460.

Tarbell, J.M. and Cancel, L.M., 'The glycocalyx and its significance in human medicine', *Journal of Internal Medicine* (2016), DOI: 10.1111/joim.12465.

Thom, M. et al., 'Seasonal biostabilization and erosion behavior of fluvial biofilms under different hydrodynamic and light conditions', *International Journal of Sediment Research* 30 (2015), pp. 273–84.

Turko, A.J. et al., 'Skeletal stiffening in an amphibious fish out of water is a response to increased body weight', *Journal of Experimental Biology* (2017), DOI: 10.1242/jeb.161638.

Tyler, M.J. et al., 'Inhibition of gastric acid secretion in the gastric brooding frog, *Rheobatrachus silus*', *Science* (1983), DOI: 10.1126/science.6573024.

Vaelli, P.M. et al., 'The skin microbiome facilitates adaptive tetrodotoxin production in poisonous newts', *eLife* (2020), DOI: 10.7554/eLife.53898.

Vági, B. et al., 'Parental care and the evolution of terrestriality in frogs', *Proceedings of the Royal Society B. Biological Sciences* (2019), DOI: 10.1098/rspb.2018.2737.

Vahey, M.D. and Fletcher, D.A., 'Influenza A virus surface proteins are organized to help penetrate host mucus', *eLife* (2019), DOI: 10.7554/eLife.43764.

Valente, R.H. et al., '*Bothrops jararaca* accessory venom gland is an ancillary source of toxins to the snake', *Journal of Proteomics* (2018), DOI: 10.1016/j.jprot.2017.12.009.

Van Deynze, A. et al., 'Nitrogen fixation in a landrace of maize is supported by a mucilage-associated diazotrophic microbiota', *PLOS Biology* (2018), DOI: 10.1371/journal.pbio.2006352.

Vogel, D. and Dussutour, A., 'Direct transfer of learned behaviour via cell fusion in non-neural organisms', *Proceedings of the Royal Society B. Biological Sciences* (2016), DOI: 10.1098/rspb.2016.2382.

Von Byern, J. et al., 'Biomechanical properties of fishing lines of the glowworm *Arachnocampa luminosa* (Diptera; Keroplatidae)', *Scientific Reports* (2019), DOI: 10.1038/s41598-019-39098-1.

Von Proschwitz, T., 'Bericht über die 46. Frühjahrstagung der Deutschen Malakozoologischen Gesellschaft in Vickleby auf der Ostseeinsel Öland (Schweden) vom 25. bis 28.Mai 2007', *Mitteilungen der Deutschen Malakozoo-logischen Gesellschaft* (2014), pp.1–12.

Wagner, C.E. et al., 'A rheological study of the association and dynamics of MUC5AC gels', *Biomacromolecules* (2017), DOI: 10.1021/acs.biomac.7b00809.

Walker, S. et al., 'Root exudation and rhizosphere biology', *Plant Physiology* (2003), DOI: 10.1104/pp.102.019661.

Waller, D. et al., 'Reversibility ofVasalgelTM male contraceptive in a rabbit model', *Basic and Clinical Andrology* (2017), DOI: 10.1186/s12610-017-0051-1.

Weiss, M.C. et al., 'The physiology and habitat of the last universal common ancestor', *Nature Microbiology* (2016), DOI: 10.1038/nmicrobiol.2016.116.

Werb, Z. and Lu, P., 'The role of stroma in tumor development', *Cancer Journal* (2016), DOI: 10.1097/PPO.0000000000000127.

Wetzel, L.A. et al., 'Predicted glycosyltransferases promote development and prevent spurious cell clumping in the choanoflagellate *S.rosetta*', *eLife* (2018), DOI: 10.7554/eLife.41482.

Wheeler, R.J. and Hyman, A.A., 'Controlling

compartmentalization by non-membrane-bound organelles', *Philosophical Transactions of the Royal Society B. Biological Sciences* (2018), DOI: 10.1098/rstb.2017.0193.

Whelan, N.V. et al., 'Ctenophore relationships and their placement as the sister group to all other animals', *nature ecology & evolution* (2017), DOI: 10.1038/s41559-017-0331-3.

Whittington, I.D. and Kearn, G., 'Hatching strategies in monogenean (Platyhelminth) parasites that facilitate host infection', *Integrative & Comparative Biology* (2011), DOI: 10.1093/icb/icr003.

Wieczorek, A.M. et al., 'Microplastic ingestion by gelatinous zooplankton may lower efficiency of the biological pump', *Environmental Science & Technology* (2019), DOI: 10.1021/acs.est.8b07174.

Wild, D. et al., 'Coral mucus functions as an energy carrier and particle trap in the reef ecosystem', *Nature* (2004), DOI: 10.1038/nature02344.

Wilks, A.M., 'Double-network gels and the toughness of terrestrial slug glue', *Journal of Experimental Biology* (2015), DOI: 10.1242/jeb.128991.

Winegard, T. et al., 'Coiling and maturation of a high-performance fibre in hagfish slime gland thread cells', *Nature Communications* (2014), DOI: 10.1038/ncomms4534.

Witten, J. et al., 'Selective permeability of mucus barriers', *Current Opinion in Biotechnology* (2018), DOI: 10.1016/j.copbio.2018.03.010.

Witten, J. and Ribbeck, K., 'The particle in the spider's web: Transport through biological hydrogels', *Nanoscale* (2018), DOI: 10.1039/c6nr09736g.

Wizen, G. and Gasith, A., 'An unprecedented role reversal: Ground beetle larvae (*Coleoptera*: *Carabidae*) lure amphibians and prey upon them', *PLOS One* (2011), DOI: 10.1371/journal.pone.0025161.

Wood, R. et al., 'Integrated records of environmental change and evolution challenge the Cambrian Explosion', *nature ecology & evolution* (2019), DOI: 10.1038/s41559-019-0821-6.

Wotton, R.S., 'The ubiquity and many roles of exopolymers

(EPS) in aquatic systems', *Scientia Marina* (2004), DOI: 10.3989/scimar.2004.68s113.

Woznica, A. et al., 'Bacterial lipids activate, synergize, and inhibit a developmental switch in choanoflagellates', *PNAS* (2016), DOI: 10.1073/pnas.1605015113.

Woznica, A. and King, N., 'Lessons from simple marine models on the bacterial regulation of eukaryotic development', *Current Opinion in Microbiology* (2018), DOI: 10.1016/j.mib.2017.12.013.

Wurl, O. et al., 'Sea surface microlayer in a changing ocean – A perspective', *Elementa. Science of the Anthropocene* (2017), DOI: 10.1525/elementa.228.

Xu, C. et al., 'Highly elastic biodegradable single-network hydrogel for cell printing', *ACS Applied Materials & Interfaces* (2018), DOI: 10.1021/acsami.8b01294.

Yang, D. et al., 'Enhanced transcriptions and translation in clay hydrogel and implications for early life evolution', *Scientific Reports* (2013), DOI: 10.1038/srep03165.

Yang, Y. et al., 'Evolution of nematode-trapping cells of predatory fungi of the *Orbiliaceae* based on evidence from rRNA-encoding DNA and multiprotein sequences', *PNAS* (2007), DOI: 10.1073/pnas.0702770104.

Yap, Y.U. and Marino, E., 'An insight into the intestinal web of mucosal immunity, microbiota, and diet in inflammation', *Frontiers in Immunology* (2018), DOI: 10.3389/fimmu.2018.02617.

York, L.M. et al., 'The holistic rhizosphere: Integrating zones, processes, and semantics in the soil influenced by roots', *Journal of Experimental Botany* (2016), DOI: 10.1093/jxb/erw108.

Yue, B., 'Biology of the extracellular matrix: An overview', *Journal of Glaucoma* (2014), DOI: 10.1097/IJG.0000000000000108.

Zaitsev, Y.P., 'Contourobionts in ocean monitoring', *Environmental Monitoring and Assessment* 1/7 (1986), pp. 31–8.

Zhang, C. et al., 'Evolving paradigms in biological carbon cycling in the ocean', *National Science Review* (2018), DOI: 10.1093/nsr/nwy074.3.

Zimstein, I. et al., 'Eukaryotic life in biofilms formed in a uranium mine', *Microbiology Open* (2020), DOI: 10.1002/mbo3.17.

BOOKS

Abbott, E.A. 1884. *Flatland: A Romance of Many Dimensions*. London: Seeley & Co.

Andersen, H.C. 2005. *Fairy Tales*. London: Penguin Classics.

Archibald, J.D. 2014. *Aristotle's Ladder, Darwin's Tree: The Evolution of Visual Metaphors for Biological Order*. New York: Columbia University Press.

Ashton, R. 2017. *One Hot Summer: Dickens, Darwin, Disraeli, and the Great Stink of 1858*. New Haven and London: Yale University Press.

Bailey, E.T. 2010. *The Sound of a Wild Snail Eating*. New York: Algonquin Books.

Bakewell, S. 2016. *At the Existentialist Café: Being, and Apricot Cocktails*. New York: Other Press.

Baumstark, K. 2015. *Der Tod und das Mädchen. Erotik, Sexualität und Sterben im deutschsprachigen Raum zwischen Spätmittelalter und Früher Neuzeit*. Berlin: LIT Verlag Münster.

Beevor, A. 2009. *D-Day: The Battle for Normandy*. London: Penguin.

Bell-Metereau, R. 'Searching for Blobby Fissures. Slime, Sexuality, and the Grotesque', in Murray, P. 2004. *Bad: Infamy, Darkness, Evil, and Slime on Screen*. New York: State of New York University Press.

Benn, G. 1998. *Sämtliche Gedichte*. Stuttgart: Klett Cotta.

Bereiter-Hahn, J., Matoltsy, G. and Richards, S. 1984. *Biology of the Integument, 1: Invertebrates*. Berlin, Heidelberg: Springer-Verlag.

Birkhead, T. 2013. *Bird Sense: What It's Like to Be a Bird*. London: Bloomsbury.

Birkhead, T. 2016. *The Most Perfect Thing: Inside (and Outside) a Bird's Egg*. London: Bloomsbury.

Boetius, A. and Boetius, H. 2011. *Das dunkle Paradies: Die Entdeckung der Tiefsee*. Munich: C. Bertelsmann Verlag.

Brain, R.M. 2015. *The Pulse of Modernism: Physiological Aesthetics in*

Fin-de-Siècle Europe. Seattle: University of Washington Press.

Brehm, A. 1927. *Brehms Tierleben, Band 23: Lurche*. Hamburg: Gutenberg-Verlag.

Briggs, J. 2013. *A Little Book about Mistletoe: Biology, Traditions, Harvest, Uses, Conservation, Control and How to Grow Your Own*. Stonehouse: Potamogeton Press.

Brunner, B. 2011. *Wie das Meer nach Hause kam: Die Erfindung des Aquariums*. Berlin: Wagenbach.

Butler, O.E. 1998. *Parable of the Talents*. London: Headline.

Von Byern, J. and Grunwald, I. (ed.). 2010. *Biological Adhesive Systems: From Nature to Technical and Medical Application*. Vienna, New York: Springer-Verlag.

Čapek, K. 2014. *War with the Newts*. London: Hachette.

Carson, R.L. 1998. *The Edge of the Sea*. Boston: Houghton Mifflin Harcourt.

Carson, R.L. 2011. *The Sea Around Us*. New York: Open Road Integrated Media.

Carson, R.L. 2011. *Under the Sea-Wind*. New York: Open Road Integrated Media.

Casanova, G. 1970. *History of My Life*. Baltimore: Johns Hopkins University Press.

Chesterton, G.K. 2014. *The Complete Works of G K. Chesterton*. Hastings: Delphi.

Chimileski, S. and Kolter, R. 2017. *Life at the Edge of Sight: A Photographic Exploration of the Microbial World*. Cambridge, Mass.: Harvard University Press.

Von Cluny, O. 1990. *The Body and Surgery in the Middle Ages*. Cambridge: Polity Press.

Coen, R. 2014. *Fu-go: The Curious History of Japan's Balloon Bomb Attack on America*. Lincoln: University of Nebraska Press.

Connor, S. 2004. *The Book of Skin*. Ithaca: Cornell University Press.

Corbin, A. 1986. *Pesthauch und Blütenduft. Eine Geschichte des Geruchs*. Berlin: Wagenbach.

Corbin, A. 1986. *The Foul and the Fragrant: Odor and the French Social Imagination*. Cambridge, Mass.: Harvard University Press.

Corfield, R. 2005. *The Silent Landscape: In the Wake of HMS Challenger 1872–1876.* London: Hodder & Stoughton.

Crampton, C. 2020. *Way to the Sea: The Forgotten Histories of the Thames Estuary*. London: Granta Books.

Crichton, M. 2012. *The Andromeda Strain*. New York: Vintage Books.

Curtis, V. 2013. *Don't Look, Don't Touch: The Science behind Revulsion*. Oxford: Oxford University Press.

Das, S. 2005. *Touch and Intimacy in First World War Literature*. Cambridge: Cambridge University Press.

Davies, P. 2010. *Eerie Silence: Are we Alone in the Universe?* London: Penguin.

Day, C.A. 2017. *Consumptive Chic: A History of Beauty, Fashion, and Disease*. London: Bloomsbury.

De Beauvoir, S. 1984. *Adieux. A Farewell to Sartre*. New York: Pantheon Books.

De Beauvoir, S. 2014. *The Second Sex*. London: Vintage Books.

Dickens, C. 1853. *Bleak House*. London, 2003: Penguin Books.

Dickens, C. 1873. *Little Dorrit*. London: Chapman & Hall.

Dipper, F. 2016. *The Marine World: A Natural History of Ocean Life*. Ithaca: Cornell University Press.

Dumas, A. 2000. *La Dame aux Camélias.* Oxford: Oxford University Press.

Dussutour, A. 2017. *Le Blob. Tout ce que vous avez toujours voulu savoir*. Paris: Éditions des Équateurs.

Dürer, A., in Thomas, H. 2013. *Conquest*. New York: Simon & Schuster.

Duve, K. 1999. *Regenroman*. Berlin: Kiepenheuer & Witsch.

Egan, T. 2006. *The Worst Hard Time*. New York: Houghton Mifflin Harcourt.

Von Engelhardt, D. and Wisskirchen, H. 2003. *Der Zauberberg: die*

Welt der Wissenschaften in Thomas Manns Roman. Stuttgart: Schattauer Verlag.

Enns, A. and Trower, S. 2013. Vibratory Modernism. London: Palgrave Macmillan.

Figes, O. 2017. *A People's Tragedy*. London: Random House.

Flemming, H., Neu, T.R. and Wingender, J. (eds). 2017. *The Perfect Slime*. London: IWA Publishing.

Forbes, P. 2014. *Dazzled and Deceived: Mimicry and Camouflage*. London and New Haven: Yale University Press.

Gerrard, N. 2019. *What Dementia Teaches Us about Love*. London: Penguin.

Gibson, S. 2015. *Animal, Vegetable, Mineral? How Eighteenth-century Science Disrupted the Natural Order.* Oxford: Oxford University Press.

Gladyshev, M.I. 2002. *Biophysics of the Surface Microlayer of Aquatic Ecosystems*. London: IWA Publishing.

Gleeson, S. 2019. *Constellations: Reflections from Life*. London: Pan Macmillan.

Goetz, T. 2014. *The Remedy: Robert Koch, Arthur Conan Doyle, and the Quest to Cure Tuberculosis*. New York: Penguin.

Golding, W. 2012. *Lord of the Flies*. London: Faber & Faber.

Goscinny, R. and Uderzo, A. (trans. Bell, A. and Hockridge, D.). 1975. *Asterix and the Golden Sickle*. Leicester: Brockhampton Press.

Grimm, J. and Grimm, W. 2015. *Grimm's Fairy Tales*. Alcester: Pook Press.

Harvell, D. 2016. *A Sea of Glass: Searching for the Blaschkas' Fragile Legacy in an Ocean at Risk*. Oakland: University of California Press.

Harvell, D. 2019. *Ocean Outbreak: Confronting the Rising Tide of Marine Disease*. Oakland: University of California Press.

Hasan, Z., Hug, A. Z. and Nussbaum, M. 2018. *The Empire of Disgust: Prejudice, Discrimination, and Policy in India and the US*. New Delhi: Oxford University Press.

Herz, R. 2012. *That's Disgusting: Unraveling the Mysteries of Repulsion*. New York: W.W. Norton & Co.

Highsmith, P. 2011. *Eleven*. London: Grove Press.

Highsmith, P. 2014. *Deep Water*. London: Virago.

Hinton, N. 1986. *The Heart of the Valley*. London: HarperCollins.

Hugo, V. 1994. *Les Misérables.* Ware: Wordsworth Editions.

Hugo, V. 2002. *The Toilers of the Sea*. London: Random House.

Hunter, G.K. 2000. *Vital Forces: The Discovery of the Molecular Basis of Life*. London: Elsevier.

Hurley, K. 1996. *The Gothic Body. Sexuality, Materialism, and Degeneration at the Fin de Siècle*. Cambridge: Cambridge University Press

Ings, S. 2016. *Stalin and the Scientists: A History of Triumph and Tragedy 1905–1953*. London: Faber & Faber.

Janzen, J. 2016. *Media, Modernity and Dynamic Plants in Early 20th Century German Culture*. Leiden, Boston: Brill.

Jones, D.W. 2012. *Howl's Moving Castle*. New York: Harper Collins.

Joshi, S.T. 2013. *I Am Providence: The Life and Times of H.P. Lovecraft*. New York: Hippocampus Press.

Junger, S. 2010. *The Perfect Storm: A True Story of Men Against the Sea*. London: HarperCollins.

Kavanagh, J. 2013. *The Girl Who Loved Camellias: The Life and Legend of Marie Duplessis*. New York: Knopf Doubleday.

Keynes, R. (ed.). 2000. *Charles Darwin's Zoology Notes & Specimen Lists from H.M S. Beagle*. Cambridge: Cambridge University Press.

King, S. 2011. *Danse Macabre*. New York: Simon & Schuster.

Klinger, L.S. (ed.). 2014. *The New Annotated H.P. Lovecraft*. New York: W.W. Norton & Co.

Knoll, A.H. 2003. *Life on a Young Planet: The First Three Billion Years of Evolution on Earth*. Princeton: Princeton University Press.

Koehl, M. 2006. *Wave-Swept Shore: The Rigors of Life on a Rocky Coast*. Berkeley: University of California Press.

Kolbert, E. 2014. *The Sixth Extinction: An Unnatural History*. London: A&C Black.

Kreitschitz, A. 'Biological Properties of Fruit and Seed Slime Envelope. How to Live, Fly, and Not Die', in Gorb, S. (ed.) 2009.

Functional Surfaces in Biology, Vol. 2. Dordrecht: Springer-Verlag.

Kruft, H. 1989. *Städte in Utopia: die Idealstadt vom 15. bis zum 18. Jahrhundert zwischen Staatsutopie und Wirklichkeit*. Munich: C.H. Beck.

Le Guin, U.K. 1989. *Dancing at the Edge of the World: Thoughts on Words, Women, Places*. New York: Grove Press.

Lehoux, D. 2017. *Creatures Born of Mud and Slime: The Wonder and Complexity of Spontaneous Generation*. Baltimore: JHU Press.

Lem, S. 2002. *Solaris*. San Diego: Harcourt.

Leuchs, J.C. 2012. *Anweisung zur Bereitung Des Tischlerleims, Der Knochengallerte Der Hausenblase, Des Vogelleims Und der Suppentafeln*. US.

Lima, M. 2014. *The Book of Trees*. New York: Princeton Architectural Press.

Löns, H. 1911. *Der zweckmäßige Meyer/Frau Döllmer*. Hanover: Sponholtz.

Love, M. 2016. *Good Vibrations: My Life as a Beach Boy*. New York: Faber & Faber.

Lovecraft, H.P. 2014. *The Complete Fiction of H.P. Lovecraft*. New York: Race Point Publishing.

Lyons, S.L. 2020. *From Cells to Organisms. Re-Envisioning Cell Theory*. Toronto: University of Toronto Press.

Mackessy, S.P. 2010. *Handbook of Venoms and Toxins of Reptiles*. Boca Raton: CRC Press.

Mann, T. 1955. *The Confessions of Felix Krull, Confidence Man*. London: Penguin.

Mann, T. 1999. *The Magic Mountain*. London: Vintage Books.

Margulis, L. and Sagan, D. 1995. *What is Life? The Eternal Enigma*. Los Angeles: University of California Press.

Marshall, M. 2020. *The Genesis Quest*. London: Weidenfeld & Nicolson.

Martin, A.J. 2017. *The Evolution Underground: Burrows, Bunkers, and the Marvellous Subterranean World Beneath our Feet*. New York: Simon & Schuster.

Matyssek, R. et al. 2010. *Biologie der Bäume: Von der Zelle zur globalen Ebene*. Stuttgart: Ulmer.

McNamara, K. 2009. *Stromatolites*. Welshpool: Western Australian Museum.

Matheson, R. 2009. *I Am Legend*. London: Gollancz.

Melville, H. 2003. *Moby Dick*. New York: Bantam Books.

Metzger, R. 2015. *Die Stadt – Vom antiken Athen bis zu den Megacitys: Eine Weltgeschichte in Geschichten*. Munich: Brandstätter.

Michelinie, D. 2018. *Venom: Planet of the Symbiotes*. New York: Marvel Entertainment.

Miodownik, M. 2018. *Liquid: The Delightful and Dangerous Substances That Flow Through Our Lives*. London: Penguin.

Moxham, R. 2001. *The Great Hedge of India*. London: Basic Books.

Nabokov, V. 1995. *The Annotated Lolita*. London: Penguin Books.

Nachtigall, W. and Wisser, A. 2013. *Bionik in Beispielen: 250 illustrierte Ansätze*. Berlin: Springer-Verlag.

Nussbaum, M.C. 2004. *Hiding from Humanity: Disgust, Shame, and the Law.* Princeton: Princeton University Press.

Nussbaum, M.C. 2018. *The Monarchy of Fear*. New York: Simon & Schuster.

O'Brian, P. 1979. *The Fortune of War*. London: Collins.

Ocean, A. 2015. *My Sexy Slime Girlfriend (Supernatural Romance)*.

Offenberger, M. 2012. *Das Ei: Ursprung allen Lebens*. Darmstadt: Primus Verlag.

Oken, L. 1843. *Lehrbuch der Naturphilosophie*. Zürich.

Onstott, T.C. 2017. *Deep Life: The Hunt for the Hidden Biology of Earth, Mars, and Beyond*. Princeton: Princeton University Press.

Oschmann, W. 2016. *Evolution der Erde*. Bern: UTB.

Pauly, D. 2010. *5 Easy Pieces: The Impact of Fisheries on Marine Ecosystems.* Washington DC: Island Press.

Peters, F. 2020. 'Uncanny Snails: Patricia Highsmith and the Allure of the Gastropods', in Ruth Heholt and Melissa Edmundson (eds), *Gothic Animals: Uncanny Otherness and the Animal With-Out*. London: Palgrave Macmillan.

Philbrick, N. 2000. *In the Heart of the Sea: The Tragedy of the Whaleship Essex*. New York: Viking.

Philbrick, N. 2006. *Mayflower: A Voyage to War.* New York: Viking.

Prager, E. 2011. *Sex, Drugs, and Sea Slime: The Oceans' Oddest Creatures and why they Matter*. Chicago: The University of Chicago Press.

Roche, C. 2008. *Feuchtgebiete*. Cologne: M. DuMont Schauberg.

Röhrlich, D. 2012. *Urmeer: Die Entstehung des Lebens*. Hamburg: Mareverlag.

Rowling, J.K. 2018. *Fantastic Beasts: The Crimes of Grindelwald – Original Screenplay*. London: Pottermore Publishing.

Sagan, C. 1980. *Broca's Brain: Reflections on the Romance of Science.* New York: Ballantine Books.

Sartre, J.-P. 2003. *Being & Nothingness.* Abingdon: Routledge.

Schickore, J. 2017. *About Method: Experimenters, Snake Writing Scientifically*. Chicago: University of Chicago Press.

Schilthuizen, M. 2014. *Darwins Peep Show: Wastierische Fortpflanzungsmetho-den über das Leben und die Evolution enthüllen.* Munich: dtv.

Schutt, B. 2017. *Cannibalism: A Perfectly Natural History.* Chapel Hill: Algonquin Books.

Sexton, A. 2001. *Transformations.* New York: Mariner Books.

Snyder, L. 2015. *Eye of the Beholder: Johannes Vermeer, Antoni van Leeuwenhoek, and the Reinvention of Seeing*. New York: W.W. Norton & Co.

Steinbeck, J. 2001. *The Grapes of Wrath*. London: Penguin.

Steinbeck, J. 2001. *The Log from the Sea of Cortez*. London: Penguin.

Stevenson, R.L. 1994. *The Strange Case of Dr Jekyll and Mr Hyde.* London: Penguin.

Stott, R. *Oyster.* 2004. London: Reaktion Books.

Strick, J.E. 2000. *Sparks of Life: Darwinism and the Victorian Debates over Spontaneous Generation*. Cambridge, Mass.: Harvard University Press.

Süskind, P. 2015. *Perfume: The Story of a Murderer.* London: Penguin.

Swift, J. 1992. *Gulliver's Travels.* Ware: Wordsworth Editions.

Syme, P. 2017. *Werner's Nomenclature of Colours.* London: Natural History Museum.

Theweleit, K. 2003. *Male Fantasies.* Minneapolis: University of Minnesota Press.

Vernadsky, V. 1998. *The Biosphere.* New York: Springer Science and Business Media.

Verne, J. 1980. *Twenty Thousand Leagues Under the Sea.* London: Octopus Books.

Vogel, S. 1988. *Life's Devices: The Physical World of Animals and Plants.* Princeton: Princeton University Press.

Vogel, S. 1994. *Life in Moving Fluids: The Physical Biology of Flow.* Princeton: Princeton University Press.

Vogel, S. 2013. *Comparative Biomechanics: Life's Physical World.* Princeton: Princeton University Press.

Walker, G. 2003. *Snowball Earth: The Story of a Maverick Scientist and His Theory of the Global Catastrophe that Spawned Life as We Know It.* London: Bloomsbury.

Walker, G. 2007. *An Ocean of Air: A Natural History of the Atmosphere.* London: Bloomsbury.

Walton, I. 1653. *The Compleat Angler, Or Contemplative Man's Recreation.* London.

Ward, P. 2009. *The Medea Hypothesis. Is Life on Earth Ultimately Self-Destructive?* Princeton: Princeton University Press.

Wells, H.G. 2006. *A Short History of the World.* London: Penguin

Wells, H.G. 2012. *The Invisible Man.* London: Penguin.

Werner, F. 2015. *Schnecken: Ein Portrait.* Berlin: Matthes & Seitz.

Willmann, R. and Voss, J. 2017. *The Art and Science of Ernst Haeckel.* Cologne: Taschen.

Wilson, A. 2003. *Beautiful Shadow: A Life of Patricia Highsmith.* London: A&C Black.

Winton, T. 2015. *Island Home: A Landscape Memoir.* Melbourne: Penguin.

Woolf, V. 2019. *A Room of One's Own*. London: Penguin Classics.

Wright, J. 2016. *A Natural History of the Hedgerow*. London: Profile Books.

Yong, E. 2016. *I Contain Multitudes: The Microbes Within Us and a Grander View of Life*. London: Random House.

LINKS

Last accessed on 18 June 2019 unless otherwise stated.

Andersen. H.C. *The Little Mermaid*. Available at: http://www.andersen.sdu.dk.

Andersen, H.C. *The Great Sea Serpent*. Available at: http://www.andersen.sdu.dk.

Astrobiology Magazine. 'The rise of slime'. {https://www.astrobio.net/climate/the-rise-of-slime/}.

Brehm, A. Sippe: *Morgenröthenthier* (Eozoon) (1887). {http://www.zeno.org/Naturwissenschaften/M/Brehm,+Alfred/Brehms+Thierleben/Niedere+Thiere/Der+Kreis+der+Urthiere/Zweite+Ordnung%3A+Vielkammerige+Wurzelf%C3%BC%C3%9Fer+oder+Foraminiferen+(Polythalamia)/Anhang/Sippe%3A+Morgenr%C3%B6thenthier+(Eozoon)}.

Carnall, M. 'Why do cephalopods produce ink? And what's ink made of, anyway?' {https://www.theguardian.com/science/2017/aug/09/why-do-cephalopods-produce-ink-and-what-on-earth-is-it-anyway}.

Casanova, G. *Erinnerungen*. {https://gutenberg. spiegel.de/buch/erinnerungen-611/12}.

Copley, Jon: 'Ooze cruise'. {https://www.newscientist.com/article/mg16522294-300-ooze-cru}.

Darwin, C. 'But if (& oh what a big if!)'. {https://www.darwinproject.ac.uk/letter/DCP-LETT-7471.xml}.

Davis, N. 'New antibiotics could be developed using fish slime, scientists say'. {https://www.theguardian.com/society/2019/mar/

31/new-antibiotics-could-be-developed-using-fish-slime-scientists-say}.

Deep Carbon Observatory. 'Life in deep earth totals 15 to 23 billion tonnes of carbon – hundreds times more than humans'. {https://deepcarbon.net/life-deep-earth-totals-15-23-billion-tonnes-carbon}.

Doyle, A.C. 'When the World Screamed'. {http://gutenberg.net.au/ebooks01/0100031h.html}.

Dürer, A. *Etliche Underricht zu Befestigung der Stett, Schloss und Flecken*. {https://www.e-rara.ch/doi/10.3931/e-rara-9248}.

Eliot, T.S. 'What Dante means to me'. {https://wordandsilence.com/2016/ 05/19/t-s-eliot-on-dante/}.

Engber, D. 'Out of slime'. {https://slate.com/technology/2016/07/ghost-busters-made-slime-a-national-obsession-can-the-reboot-make-ooze-cool-again.html}.

Fardin, M. 'Answering the question that won me the Ig Nobel prize: Are cats liquid?' {https://theconversation.com/ answering-the-question-that-won-me-the-ig-nobel-prize-are-cats-liquid-86589}.

Geyer, W.R. 'Where the rivers meet the sea'. Available at: https://www.whoi.edu/oceanus/feature/where-the-rivers-meet-the-sea/.

GoJelly. 'H2020 EU project'. {https://gojelly.eu/}.

Golding, W. 'Nobel lecture'. {https://www.nobelprize.org/prizes/literature/1983/golding/ lecture/}.

Guiterman, A. 'Ode to the Amoeba'. {https://www.newyorker.com/magazine/1932/11/05/ode-to-the-amoeba}.

Hendrix, S. 'He always hated women. Then he decided to kill them'. {https://www.washingtonpost.com/graphics/2019/local/yoga-shooting-incel-attack-fueled-by-male-supremacy/?utm_term=.04e03bdf65c5}.

Henriques, M. 'The idea that life began as clay crystals is 50 years old'. {http://www.bbc.com/earth/story/20160823-the-idea-that-life-began-as-clay-crystals-is-50-years-old}.

Huxley, Thomas. 'On the physical basis of life'. {https://mathcs.clarku.edu/huxley/CE1/PhysB.html}.

Jakobsen, H. 'Nasty nasal parasite'. {http://science-nordic.com/nasty-nasal-parasite}.

Jones, J. 'Edvard Munch: *Love and Angst* review'. {https://www.theguardian.com/artanddesign/2019/apr/09/scream-edvard-munch-love-and-angst-review-british-museum}.

Juler, E. 'Life forms: Henry Moore, morphology and biologism in the interwar years'. {https://www.tate.org.uk/art/research-publications/henry-moore/edward-juler-life-forms-henry-moore-morphology-and-biologism-in-the-interwar-years-r1151314}.

Knauer, R. 'Gut leben im Endlager'. Available at: https://www.spektrum.de/news/bakterien-koennten-endlager-beeintraechtigen/1433500.

Lobe, A. 'Der Netflix-Algorithmus macht Kunst berechenbar'. {https://www.sueddeutsche.de/medien/netflix-algorithmus-daten-house-of-cards-1.4280852}.

Mapp, K. 'US Navy synthetically recreates bio-material to assist military personnel'. {https://www.navy.mil/submit/display.asp?story_id=98521}.

Marshall, M. 'How the first life on Earth survived its biggest threat – water'. Available at: https://www.nature.com/articles/d41586-020-03461-4.

Nussbaum, M. 'The roots of male rage, on show at the Kavanaugh hearing'. {https://www.washingtonpost.com/news/democracy-post/wp/2018/09/29/the-roots-of-male-rage-on-show-at-the-kavanaugh-hearing/?utm_ term=.d384091544b1}.

Onion, R. 'Ode to green slime'. {https://www.theatlantic.com/technology/ archive/2015/02/ode-to-green-slime/385088/#}.

Powell, K. 'The super-hero in your vagina'. {https://mosaicscience.com/story/bacterial-vaginosis/}.

Robinson, P. 'Under the Sun/The Man-Eating Tree'. {https://en.wikisource.org/wiki/Under_the_Sun/The_Man-Eating_Tree}

Rodwell, I. 'Hedge'. {https://liminalnarratives.com/2018/06/17/hedge/}.

Rosen, L.D. 'Phantom pocket vibration syndrome'. {https://www.psychology-today.com/intl/blog/rewired-the-psychology-technology/201305/phantom-pocket-vibration-syndrome}.

Rupp, R. 'The luke-warm, gluey, history of portable soup'. {https://www.nationalgeographic.com/people-and-culture/food/the-plate/2014/09/25/the-luke-warm-gluey-history-of-portable-soup/}.

Starkey, N. 'Blobfish voted world's ugliest animal'. {https://www.theguardian.com/environment/2013/sep/12/blobfish-world-ugliest-animal}.

Swift, J. 'The Lady's Dressing Room'. https://www.poetryfoundation.org/poems/50579/the-ladys-dressing-room

Verne, J. *Twenty Thousand Leagues under the Sea*. {https://www.gutenberg.org/files/2488/2488-h/2488-h.htm}.

Watson, T. 'These bizarre ancient species are rewriting animal evolution'. Available at: https://www.nature.com/articles/d41586-020-02985-z.

White, R. 'Mr Micawber the hermit crab'. {https://thecabinetofcuriosity.net/2018/03/11/mr-micawber-the-hermit-crab/}.

Acknowledgements

This book has many patrons and I thank each of them. My mother made sure I couldn't imagine a world without books and if it weren't for my father, I may never have thought about writing one myself. He might have also given me a love of the obscure, which might account for the choice of topic, even if his preference was a book on sticklebacks.

Sorry, Dad, that they didn't make it into the book, but I can at least report on their most spectacular slime: males build the nest and use a unique slime as glue. It's called spiggin and it is secreted by their enlarged kidneys, which don't provide any other vital functions during that time – just great dads all around. The road from concept to readable manuscript is long, but I had the support of many friends, researchers and friends in research who shared words of encouragement, advice, expertise and constructive criticism on the manuscript. I would like to thank all of them for their support: Petra Ahne, Inti Reiland, Christian Schuberth, Stefan Luschnig, Annette Atzpodien, Petra Dersch, Timo Betz, Anita Tselikas, Hans-Curt Flemming, Bettina Weber, Thorsten Becker, Dietmar Müller, Oliver Wurl, Katharina Ribbeck, Christian Wild, Janek von Byern, Mark Denny, Oliver Lieleg, Dieter Braun, Moritz Thom, Sandy Winston, Becky Rinehart, Robert Michael Brain, Ian Rodwell, John Faithfull, Linda Campbell, Clare Graham, Valerie Curtis, Philip Hoare, Hildegard Schulze, Antje Bachmann, Zena Werb, Morten Iversen and Antje Boetius. I'm also lucky enough to have in-laws who were happy to turn their infinite patience to studying and commenting on my half-written manuscript – staying up half the night reading with a newborn grandchild in their arms when necessary.

Any remaining errata are naturally the fault of no one other than myself. I'd also like to thank everyone at my German publisher, Matthias & Seitz Berlin, not least those who played a part behind the scenes. My gratitude goes to Granta UK as well for the English version of the book, and Ayça Türkoğlu for her outstanding translation. Never have my words read better. Many thanks go to Laura Barber, my editor and – even in these pandemic times – an endlessly enthusiastic and always encouraging partner in slime. It's been such a pleasure.

As has been each and every interaction with Christine Lo and Linden Lawson. I could not have wished for a better slime family.

This book has been a real joy, but a marathon too – with plenty of tough slogs over the years that I would have struggled to get through without my husband and our two sons. If they ever grew tired of my talking about slime, they never mentioned it. If they ever had plans of taking a family holiday that didn't revolve around interesting slimes and a writing desk with a good view, I never knew. And I can't thank them enough. My husband accompanied me every step of the way – his enthusiasm never waned. The one thing he's never shared, though, has been my love of Jane Austen. It's a shame because, in my experience, it is a truth universally acknowledged that a writer with a family and a book project on the go must be in want of a partner like him.

Index